Ingenieurwissenschaftliche Bibliothek
Engineering Science Library
Herausgeber / Editor: István Szabó, Berlin

István Gyarmati

Non-equilibrium Thermodynamics

Field Theory and Variational Principles

Springer-Verlag
Berlin Heidelberg GmbH 1970

Dr. István Gyarmati

Professor of Physics at the University of Gödöllő, Hungary

Originally Published as:

NEMEGYENSULYI TERMODINAMIKA

Müszaki Könyvkiadó (Technical Publishers), Budapest, 1967

Translated from the Hungarian

by Eve Gyarmati and Wolfgang F. Heinz

ISBN 978-3-642-51069-4 ISBN 978-3-642-51067-0 (eBook)

DOI 10.1007/978-3-642-51067-0

With 6 Figures

Library of Congress Catalog

Card Number 70-979990.

Title No. 4252

Dedicated to

LARS ONSAGER

Foreword by the Editor

True thermodynamics, as distinct from equilibrium or reversible thermodynamics, (also called non-equilibrium or irreversible thermodynamics) is one of the newest disciplines of physics. The theoretical principles were published by LARS ONSAGER in "The Physical Review" in 1931, a time when quantum mechanics, for instance, was already a well-defined formal discipline. Non-equilibrium thermodynamics is the universal theory of macroscopic processes and, as these processes are irreversible, it is hardly necessary to stress its importance, both theoretical and practical. A year ago LARS ONSAGER was awarded the Nobel Prize for chemistry in recognition of his establishment of this discipline. These facts more than justify the inclusion of the work of Professor GYARMATI, founder of the modern Hungarian school of thermodynamics, in the "Engineering Science Library".

The importance of this work within the international technical literature can best be appreciated by reference to recent developments in thermodynamics. Following Onsager's fundamental work, little progress was made in this field up to the end of World War II. Subsequently the German school of J. MEIXNER, the Belgian school of I. PRIGOGINE, the Dutch school of S. R. de GROOT and the Hungarian school of I. GYARMATI developed the theory and applied it on a large scale. As a result of intensive research, the theory seemed to be substantially complete about 1960, although it lacked the uniformity and completeness characteristic of the mechanics of continua and electrodynamics. Specifically, the pertinent publications revealed the absence of a unified field theory and of a universal variational principle which, like Hamilton's theory for reversible processes, would round off the existing theory. The first shortcoming is immediately apparent if we consider the brilliant completion of the mechanics of continua by C. TRUESDELL and W. NOLL. As for the variational principle, it is sufficient to recall the numerous experiments carried out since about 1964 in connection with the generalization of the minimum principle of entropy production, valid for steady-state processes only.

The English edition of Professor Gyarmati's work, substantially unchanged from the original Hungarian edition, is presented in the hope that it will convince the reader that the Hungarian school of thermodynamics has made an important contribution to the elimination of the above-mentioned shortcomings.

Berlin, Autumn 1969

István Szabó

Preface to the English Edition

Although nearly three years have elapsed since the publication of this work in Hungarian, it was decided to publish the English edition in the same form as the original, apart from some minor modifications. Since, recent research has been directed to the development of an exact theory of non-linear irreversible processes; we suggest to readers interested in similar tasks — such as the continuation of this book — that they should study some new publications: "On the most general form of the Thermodynamic Integral Principle", Z. phys. Chem. 239 (1968) 133, and particularly: "On the Governing Principle of Dissipative Processes", Ann. Phys. 7 (1969) 23.

I have to thank my wife and Mr. W.F. Heinz for the translation of the very concise Hungarian text. I also wish to express my gratitude to Dr. L. Karádi and Mr. Gy. Vincze for reading the typescript with such care and to Mrs. A. Röszler, who typed the manuscript with great patience.

I am deeply indebted to Professor István Szabó for making this edition available so quickly and for including my work in the "Engineering Science Library". Finally, I would like to express my thanks to Springer-Verlag for the excellent edition and to the editorial staff for their readiness to meet my wishes.

Gödöllő, Summer 1969

István Gyarmati

Abbreviated Preface of the Hungarian Edition

Non-equilibrium (irreversible) thermodynamics is a relatively new science whose fundamentals were elaborated by LARS ONSAGER in 1931, in his famous works considered today as classical already. This theory, similarly to its classical predecessor (equilibrium or reversible thermodynamics), is primarily of a phenomenological character, though the so-called Onsager reciprocal relations, which are pillars of the theory, have been confirmed hitherto rather by statistical methods making use of the hypothesis of microscopic reversibility. In this book only the macroscopic fundamentals of the theory are discussed. This development of non-equilibrium thermodynamics is similar to the exact treatment of classical field theory (continuum mechanics, electrodynamics), by which the uniform description of mechanical, electromagnetic and thermal phenomena is made possible. Consequently it can be stated that the results of non-equilibrium thermodynamics must be taken into consideration by the physicist, physico-chemist, energeticist, plasma researcher, the chemical engineer dealing with the operations of the chemical industries, reaction kineticist as well as by the biophysicist and biologist. The widespread range of applicability of irreversible thermodynamics results from the fact that *in nature all macroscopic processes are irreversible.*

The first part of this book is based on the fundamentals of the classical field theories and the non-equilibrium thermodynamics is treated in an inductive way. In the second part of the book the variational principles of thermodynamics are developed and the fundamental equations of the theory are derived from a new integral principle. The material of the second part—disregarding some parts of Chapters IV and V—has hitherto not been published in the form of a book. This particularly refers to Chapters IV and VI, several results of which (in fact, in case of Chapter VI all the results) are the products of research carried out in recent years in the Department for Physical Chemistry of the Polytechnical University of Budapest.

I want to express gratitude to Professors Dr. G. SCHAY, Dr. V. K. SEMENCHENKO, Dr. J. P. TERLETSKY and Dr. G. VOJTA for their remarks concerning several parts of the book and particularly for having suggested the idea of writing this book.

I am indebted to my coworkers, particularly Dr. J. VERHÁS, Dr. H. FARKAS and Dr. Sz. BÖRÖCZ for the careful critical reading of the manuscript and for many helpful suggestions.

Budapest, September 1966

István Gyarmati

Contents

Second Part

Variational Principles

Appendix

Introduction

The matter which continuously occupies the universe in macroscopic size has from physical viewpoints mechanical, electromagnetic and thermal properties and forms of motion. The simultaneous description of all these properties and their variation in space-time is of fundamental importance in theoretical and practical respects. For this reason a researcher investigating laws of different state variations of matter cannot solely rely upon the knowledge of laws referring to approximative system models concerned with one of the three enumerated forms of motion of macroscopic substances only. Indeed, in very extreme cases only or at practically seldom fulfilled conditions several properties or forms of motion of the material systems can be disregarded. Nowadays, macroscopic physics, specialized to mechanics, electrodynamics and thermodynamics, can scarcely cope with the tasks raised by every domain of the rapidly developing technical sciences. At present the development of a unified theory is required mainly by technical sciences and industries in order to describe simultaneously all the physical, chemical, etc. forms of motion of macroscopic matter. From the viewpoint of the physicist this practical requirement means that the development of a unified theory of the classical field theories, based upon the continuity axioms, is becoming urgent.

However, it is wellknown that a unified representation of the classical field theories—continuum mechanics, electrodynamics and thermodynamics—cannot be given today, since a suitably clear and precise formulation of the set of axioms of thermodynamics is not yet available. Thus, the development of a model unifying all mentioned phenomena in the way of the classical field theories will be promising only if thermodynamics has been presented in accordance with the viewpoints of field theory as well, and, further if the theory thus developed can be all-embraced by a few axioms or by one single variational principle.

It has already been emphasized by several authors that thermodynamics (similarly to the mechanics and electrodynamics of continua) can be considered as a typical field theory. When using this conception —beyond the wellknown postulates of the classical field theory—it must be assumed that the mass or volume elements (cells) of a continuous medium can be considered as equilibrium systems, whose states can be characterized by the equilibrium state parameters despite the fact that various processes take place between the neighbouring cells. As a consequence of this condition—which is usually called local or cellular equilibrium—the non-equilibrium states of a continuous medium can be described by scalar, vector and tensor fields of macroscopic state parameters, which depend, in general, on space-time. Thus in thermodynamics, in a way similar to hydro- and electrodynamics, state variations are governed by partial differential equations.

In order to treat the state parameters of thermostatics as field quantities, the fundamental laws of the equilibrium theory must be rewritten in the form of local equations. However, in these local equations variables of mechanical and electrodynamic character are included also, for which the field equations, wellknown from mechanics and electrodynamics, hold. Thus, the unambigous introduction of thermodynamic variables—occuring in non-local fundamental laws of equilibrium heat theory—into the general field theory, is not an easy task. Anyway, at present the development of thermodynamics as a field theory as a unified continuum theory, is a current and by far not finished problem in physics.

The aim of this work, apart from its summarizing character, is to contribute to the solution of many fundamental problems. This, however, can be achieved only by presenting thermodynamics taking into consideration the fundamental postulations of classical field theories. Accordingly, the most relevant fundamentals of the classical field theories are outlined in Chapter I.

In Chapter II, the total set of balance equations of the classical field theories is dealt with, both in local and in substantial forms. In this part, particular attention has been given to the derivation of the balance equations of multicomponent superposed media.

In Chapter III, the most general description of the Onsager non-equilibrium (irreversible) thermodynamics is given, at least for multicomponent and reacting hydro-thermodynamic systems. After the local formulation of the First and Second Laws the entropy balance equation, playing a central role in the theory of irreversible processes, is determined. This is followed by the overall development and analysis of Onsager's linear laws and the set of reciprocal relations for the case of anisotropic and isotropic materials of any kind.

In Chapter IV, the basic principles and logical fundamentals of the general theory are given. The fundamental properties of the non-equilibrium potentials (dissipation functions) are discussed. The variational principle (the principle of least dissipation of energy) which is equivalent to the entire Onsager theory is formulated in a general form corresponding to the spirit of field theories. Alternative forms of the variational principle are given which proved to be very fruitful both in theoretical and practical respects.

In Chapter V, the principle of minimum production of entropy is described and the relation of this principle with the Onsager principle is elucidated. It is demonstrated that the principle of the minimum production of entropy is not a new variational principle independent of the Onsager principle, but only an alternative reformulation of the latter valid for stationary states. With the aid of forms of the variational principle valid for stationary cases the general theory of stationary states as well as the classification of the stationary systems is given. The general conditions of stationarity is determined for the case of heat conduction, multi-component diffusion, chemical reactions, etc. and the stability of such states is examined.

In Chapter VI, the Fourier equation of heat conduction (in every possible picture), the total set of Fick's equations of multi-component isothermal diffusion, the generalized Navier-Stokes equation of viscous flow are derived from the new (force) representation of the variational principle of the least dissipation of energy. It is proved by the derivation of the above-mentioned equations from the new (force) representation of the principle of least dissipation of energy that this representation is more fruitful than the original one. Furthermore, by starting from the new representation we have a possibility to formulate a new integral principle of thermodynamics. After generally formulating the integral principle and the Lagrange function of thermodynamics it is shown that the Euler-Lagrange equations belonging to the integral principle are equivalent to the total set of transport equations. As a direct application of the integral principle, the derivation of transport equations governing different non-isothermal phenomena with cross-effects, is dealt with. The relations between the integral principle of thermodynamics and the form of the Hamilton principle for fields is discussed. Finally, after the determination of the canonical field equations referring to the integral principle of thermodynamics, the Legendre transformations of the dissipative Lagrange and Hamilton densities are treated and the canonical form of the dissipative integral is given.

In the Appendix, a short summary of the employed mathematical formalism is to be found. It is hoped that hereby the understanding of the used vector, tensor etc. symbolism and operations will be facilitated.

First Part

Field Theory and Thermodynamics

I. Fundamental Concepts of the Field Theories

It is our main task to formulate the fundamentals of thermodynamics in a manner adequate to continuous models of physical systems, and consequently the knowledge of the basic concepts of the theory of fields required by such system models is relevant. In this introductory chapter, our program is to outline briefly the basic concepts and theorems of the classical field theories which will be important for the understanding of subsequent chapters. Though we are convinced that the fundamentals necessary for the continuous understanding of the subject are contained in this Chapter, we draw the attention to the excellent monographs by TRUESDELL-TOUPIN [1], and SEDOV [2], in which the systematic and exact development of classical field theories can be found.

1. The Aim of Classical Field Theory. Deformation

The general task of the classical field theory is the development of mechanics, electrodynamics and thermodynamics of continuous systems in the three-dimensional Euclidean space[1]. More particularly, the main aim of the classical field theory is the investigation of those partial differential equations, which are valid in the Euclidean space for the mechanical, electromagnetic and thermal state parameters depending on space and time in such a way that the evolution of the state parameters is given by them as space-time functions. The assumption of Euclidean space is a characteristic of classical field theories. Hereby the introduction of Cartesian systems of coordinates is always possible.

---

[1] Of course, a more general task of the classical field theory is also the invariant description of the above-mentioned theories satisfying also requirements of the theory of relativity in the Minkowskian or Riemannian space. However, in this work, presentation of the basic equations in an invariant form is omitted.

If $\boldsymbol{i}$, $\boldsymbol{j}$, and $\boldsymbol{k}$ are the unit vectors of a rectangular Cartesian references system, the position of two arbitrary points in this common frame is determined by the vector

$$\boldsymbol{R} = X_1\boldsymbol{i} + X_2\boldsymbol{j} + X_3\boldsymbol{k} \tag{1.1a}$$

and

$$\boldsymbol{r} = x_1\boldsymbol{i} + x_2\boldsymbol{j} + x_3\boldsymbol{k} \tag{1.1b}$$

where X_1, X_2, X_3 and x_1, x_2, x_3 are the rectangular coordinates in the common frame of the points in question.

Let us now construct the mathematical expressions describing the deformation of a material medium in the physical sense. Let V and V^* be two regions of space containing a portion of matter of continuous distribution, and let us consider the deformation of the matter present in the region V into the region V^* (Fig. 1). Let $\boldsymbol{R}$ be the vector of an

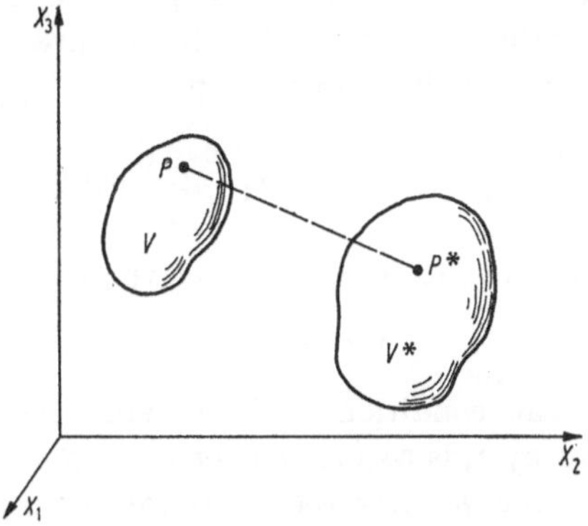

Fig. 1

arbitrary point P of the region V, which, during the deformation, is carried to the point P^* of vector $\boldsymbol{r}$ of the region V^*. This deformation—and its inverse—is described by the transformations

$$\boldsymbol{r} = \boldsymbol{r}(\boldsymbol{R}) \tag{1.2a}$$

and

$$\boldsymbol{R} = \boldsymbol{R}(\boldsymbol{r}). \tag{1.2b}$$

While the vector $\boldsymbol{R}$ runs over a set of points of the region V, its image $\boldsymbol{r}$ in the region V^* runs over a set of points of the region V^*. The relation is, of course, symmetric and thus it can also be said that the region V is deformed into V^*, or according to the inverse function, V^* is deformed into V.

2. Continuity

In the classical field theory the continuity axiom, expressing the infinite divisibility of space is used for the general characterization of the models of physical fields continuously occupied by matter.

Axiom of continuity: *Any deformation can be described only by single-valued transformations which can be continuously differentiated as many times as required* [1].

The continuity axiom excludes every physically irreal deformation, moreover, often physically permissible singularities as well. Such singularities of the classical field theories built upon the continuity axiom are distributed either on points, curves or surfaces, i.e. they are isolated singularities. The treatment of these *isolated singularities*, at least in general, does not belong to the field theory and requires separate attention. In other cases, it is sufficient to require the continuous differentiability quasi-everywhere. This condition is weaker than the axiom of continuity, however, in the majority of practical cases it is sufficient. This condition mathematically requires that the absolute scalar of the Jacobian of the deformation

$$J \equiv \left| \frac{\partial(x_1, x_2, x_3)}{\partial(X_1, X_2, X_3)} \right| = \left| \det\left(\frac{\partial x_\beta}{\partial X_\alpha}\right) \right| (\alpha, \beta = 1, 2, 3) \tag{1.3}$$

does not vanish or becomes infinite, hence the condition

$$0 < J < \infty \tag{1.4}$$

must be fulfilled. This conditions is often called the *permanence axiom* of matter, because by this axiom it is ensured *that a portion of matter present in a finite and positive volume cannot be deformed into one of zero or infinite volume.* Since for the volume elements of the deformed and undeformed medium the relation

$$\mathrm{d}V^* = J\,\mathrm{d}V \tag{1.5}$$

is valid, if and only if the permanence axiom (1.4) is satisfied, thus (1.5) is equivalent to (1.4), i.e. to the permanence of matter. Owing to the fact that (1.5) is sufficient for the formulation of the material equation of continuity, it can be stated that in field theories relying upon the continuity equations, the axiom of the permanence of matter (1.4) is of basic importance.

3. Motion

The motion of a continuous medium (deformation of elastic or plastic bodies, flow of liquids and gases etc.) is mathematically represented by a continuous transformation of the three-dimensional

Euclidean space into itself. If the motion is described in a rectangular Cartesian system, both the transformations

$$\boldsymbol{r} = \boldsymbol{r}(\boldsymbol{R}, t) \tag{1.6a}$$

and the inverse

$$\boldsymbol{R} = \boldsymbol{R}(\boldsymbol{r}, t) \tag{1.6b}$$

are motions, if the real parameter t in the interval

$$-\infty < t < \infty \tag{1.7}$$

is identified with astronomical time. The choice of the initial moment $t = 0$ is arbitrary for the examination of any actual motion. On the other hand, the transformation (1.6) describing the motion represents a deformation of the type (1.2) for each fixed time $t' = C$. Hence, *motion is a one-parameter (time parameter) family of deformations.*

Let us give the axiom of continuity with respect to time: *Any transformation expressing a motion can be continuously and partially differentiated with respect to time as many times as needed* [1].

Practically, the continuous differentiability of motions (1.6) up to the third order with respect to time is sufficient because, by this condition—though weaker as the axiom of continuity—the existence and the continuity of the acceleration functions, fundamental in dynamics, are already ensured.

It is a result of the axioms of continuity that during the motion of a continuous medium, a region is always transformed into a region, a surface into a surface, and a curve into a curve. It also follows from the axioms of continuity that if a portion of matter contained in an infinitesimal volume element of a continuous medium is interpreted as a "particle"[1], the separate "particles" undergoing a motion according to (1.6) always remain separate ones and keep their individuality during the motion. It is evident that every pathological (physically irreal) motion is excluded by the continuity axioms. However, it is also known that simultaneously with the elimination of pathological motions, many physically real motions are excluded from the field theories relying upon the continuity axioms, such as e.g. collision, impact, rolling, sliding, shocks, etc. In general, the description of the motion of isolated singularities i.e. points, curves and surface is not the task of field theories based on the continuity axioms. In other cases, if singularities are not isolated (for instance, if they do not strictly refer to points, curves, surfaces or to instants), but refer to so-called "smooth models", the basic equations valid for such models can be derived from the basic equations of the field theories as a limiting case. In fact, this is the reason why we are concerned only with the treatment of thermodynamics of continuous media, because the fundamental equations of the so-called "non-continuous" (discontinuous) system models

[1] This fictive "particle" of the classical field theory must not be identified with the particles of the corpuscular theories, for instance: molecules, ions, atoms, elementary particles.

consisting of homogeneous sub-systems are obtained as limiting cases from the fundamental equations of the field theories.[1]

4. Material and Spatial Description

In classical field theories two different methods of description are applied for the treatment of kinematics of continuous media, the roots of which originate from the Lagrangian and Eulerian description of hydrodynamics. The Lagrangian description is closer to the methods of point mechanics, whereas the viewpoints of field theory are better emphasized by the Euler method. In order to differentiate clearly between the two types of treatment let us agree to the following:

1. The coordinates X_1, X_2, X_3, of $\boldsymbol{R}$ are associated to a material point ("particle") of the continuous medium.

2. The coordinates x_1, x_2, x_3 of $\boldsymbol{r}$ represent the points of the Euclidean space which is occupied by the continuous matter.

In our convention according to the motion (1.6a) the coordinates X_1, X_2, X_3, linked to the selected "particle" at an arbitrary moment t, are invariant in time. Hence, the "particle" is represented by the X_1, X_2, X_3 coordinate triple, i.e. the coordinates X_1, X_2, X_3 can be considered as the "name" of the "particle". The place of the "particle" provided in this way by the "name" X_1, X_2, X_3 or $\boldsymbol{R}$ is determined in space according to (1.6a) by $\boldsymbol{r}$ in any arbitrary moment t. Briefly, the places occupied in the course of time by the "particle" $\boldsymbol{R}$ are registered by the motion (1.6a), and consequently the path of a "particle" is obtained from it if $\boldsymbol{R}$ is constant. Similarly, by the inverse transformation (1.6b) it is shown, which "particle" $\boldsymbol{R}$ is found at the place $\boldsymbol{r}$ of space in the moment t. Assigning different values to the X_1, X_2, X_3 coordinate triple at a fixed moment t', by transformation (1.6a) give, the position of different "particles" at the fixed time.

The Lagrange and Euler methods differ from each other in the variables considered as independent fundamental ones while describing the actual problems of motion.

The description is of the Lagrange type if the variables $\boldsymbol{R}$ or $\{X_1, X_2, X_3\}$ and t are considered as independent basic ones. In this case the equation of motion used in the dynamics of mass points is obtained for the dynamical characterization of the motion of individual "particles". This is the *material* description, often also called *substantial,* since in this case the reference system moves together with the continuum.

---

[1] Cf. for instance, in case of thermodynamics the excellent monography of de Groot and Mazur [3].

The description is of the Euler type if the variables $\boldsymbol{r}$ or $\{x_1, x_2, x_3\}$ and t are considered as independent basic ones. In this case the motion of the continuous medium is determined with respect to a system of coordinates fixed in space. This desription is frequently called *spatial* description, e.g. the velocity and acceleration fields of the moving continuum are often presented in this way.

Let us define the velocity of a "particle" $\boldsymbol{R}$ in the Euclidean space. Since $\boldsymbol{R}$ has to be constant in the course of the motion (1.6a) the velocity of the selected "particle" is determined by the partial derivative of $\boldsymbol{r}$ with respect to time. Hence, by definition

$$\boldsymbol{v} \equiv \left(\frac{\partial \boldsymbol{r}}{\partial t}\right)_{\boldsymbol{R}} \tag{1.8}$$

is the velocity of the $\boldsymbol{R}$ "particle", and consequently we have from (1.6a)

$$\boldsymbol{v} = \boldsymbol{v}(\boldsymbol{R}, t). \tag{1.9}$$

In general, the velocity of the "particle" $\boldsymbol{R}$ is a function of time. Consequently, the concept of velocity is related to the "particle", and thus primarily the Lagrangian (material) description is required by the velocity definition (1.8).

The spatial description is obtained if $\boldsymbol{R}$ is eliminated from (1.9) by substituting the inverse motion (1.6b):

$$\boldsymbol{v} = \boldsymbol{v}(\boldsymbol{r}, t). \tag{1.10}$$

According to this relation, velocity is obtained as a function of time for a given place. Hence, the Eulerian velocity field of the moving continuum is represented by (1.10).

In the theory of fields it is often neccessary to give the relations between material and spatial descriptions. This means that procedures are needed to unify the two methods and which allow to follow the motions of the individual "particles" of the continuous medium and at the same time describe the variations of the velocity field as well. In the following, the formulae connecting the two descriptions are given in an elementary way.[1]

Let us assume, that certain scalar, vector and tensor quantities belong as state parameters to each point of the continuum. These are space-time functions, such as the density of matter: $\varrho = \varrho(\boldsymbol{r}, t)$, the density of electric charge: $\varrho_e = \varrho_e(\boldsymbol{r}, t)$, pressure: $p = p(\boldsymbol{r}, t)$, temperature: $T = T(\boldsymbol{r}, t)$, velocity: $\boldsymbol{v} = \boldsymbol{v}(\boldsymbol{r}, t)$, the pressure tensor $\mathbf{P} = \mathbf{P}(\boldsymbol{r}, t)$, etc. All these properties — which are mathematically space-

[1] In case of an exact and general treatment of the field theory, the use of the theory of double tensor fields is required, which will be omitted here. Cf. page 337 [1].

time functions—are called *field quantities*. Let A be a scalar of a field quantity of an arbitrary tensorial order (hence A may be a scalar or a component of a vector or tensor of arbitrary order), and let us analyse its variation in the course of time both in material and spatial descriptions. Differentiating the function

$$A = A(\boldsymbol{R}, t) \tag{1.11a}$$

and

$$A = A(\boldsymbol{r}, t) \tag{1.11b}$$

corresponding to the two descriptions with respect to time and considering the motion (1.6), we have the following expressions:

$$\frac{\mathrm{d}A}{\mathrm{d}t} = \left(\frac{\partial A}{\partial t}\right)_{\boldsymbol{R}} + \frac{\partial A}{\partial \boldsymbol{R}} \cdot \frac{\mathrm{d}\boldsymbol{R}}{\mathrm{d}t} \tag{1.12a}$$

and

$$\frac{\mathrm{d}A}{\mathrm{d}t} = \left(\frac{\partial A}{\partial t}\right)_{\boldsymbol{r}} + \frac{\partial A}{\partial \boldsymbol{r}} \cdot \frac{\mathrm{d}\boldsymbol{r}}{\mathrm{d}t} \tag{1.12b}$$

which can be interpreted in connection with a particular case.

In the sense of our previous convention, the coordinate X_1, X_2, X_3 are data of such a frame of reference, which moves together with the "particle". Since the place of the "particle" does not change in this frame of reference, the relations $\frac{\mathrm{d}\boldsymbol{R}}{\mathrm{d}t} = 0$ and owing to (1.6a) $\frac{\mathrm{d}\boldsymbol{r}}{\mathrm{d}t} = \left(\frac{\partial \boldsymbol{r}}{\partial t}\right)_{\boldsymbol{R}}$ are valid, which yield from (1.12) the following forms:

$$\frac{\mathrm{d}A}{\mathrm{d}t} = \left(\frac{\partial A}{\partial t}\right)_{\boldsymbol{R}} \equiv \frac{\mathrm{D}A}{\mathrm{D}t} \tag{1.13a}$$

$$\frac{\mathrm{d}A}{\mathrm{d}t} = \left(\frac{\partial A}{\partial t}\right)_{\boldsymbol{r}} + \frac{\partial A}{\partial \boldsymbol{r}} \cdot \left(\frac{\partial \boldsymbol{r}}{\partial t}\right)_{\boldsymbol{R}}. \tag{1.13b}$$

The time derivative in (1.13a), which is produced by the differential operator $\frac{\mathrm{D}}{\mathrm{D}t}$ and introduced in the discussed sense, is called the *material* or *substantial time derivative*. It can easily be seen that the change of any field quantity is given by the substantial time derivative with respect to a frame of reference which moves together with the "particle". If in (1.13a) the quantity A is identical with the components of the vector $\boldsymbol{r}$, we obtain exactly the velocity $\boldsymbol{v}$ of the "particle" defined in (1.8). However, the "particle" of the classical field theory present at the place $\boldsymbol{r}$ is a mass element whose barycenter is at the end point of vector $\boldsymbol{r}$. Accordingly, it can also be said that the velocity $\boldsymbol{v}$ of the "particle" is the barycentric velocity of the mass element. It is also shown by (1.13a) that in the system of coordinates X_1, X_2, X_3, fixed to the "particle", the total derivative with respect to time of a field quantity A corresponds to the substantial time derivative.

Let us analyse formula (1.13b) which represents the spatial description. From this and (1.8) we obtain the total variation of the A field quantity in case of spatial description viz.

$$\frac{\mathrm{d}A}{\mathrm{d}t} = \left(\frac{\partial A}{\partial t}\right)_{\boldsymbol{r}} + \boldsymbol{v} \cdot \nabla A, \tag{1.14}$$

i.e. the sum of the local variation $\left(\frac{\partial A}{\partial t}\right)_{\boldsymbol{r}}$ at the place $\boldsymbol{r}$ and the convective variation $\boldsymbol{v} \cdot \nabla A$ due to the displacement of the mass element present at this point of space with the barycentric velocity $\boldsymbol{v}$. It is shown by this relation that the partial (local) time derivative taken at point $\boldsymbol{r}$ is equal to the total variation in time of the field quantity A, if the centre of mass of the "particle" present at the end point of $\boldsymbol{r}$ is at rest with respect to coordinate axes x_1, x_2, x_3 fixed to the space. Comparing (1.14) with (1.13b) it follows that, in general, the total variation in time of any field quantity can be calculated with the operator equation

$$\frac{\mathrm{d}\ldots}{\mathrm{d}t} = \frac{\partial\ldots}{\partial t} + \boldsymbol{v} \cdot \nabla\ldots \tag{1.15}$$

if the substantial time derivative is understood by operator $\frac{\mathrm{d}}{\mathrm{d}t}$. It should be noted that some authors (see, for instance [4]) prefer the use of the symbol $\frac{\mathrm{D}}{\mathrm{D}t}$ to denote the substantial derivative, indicating hereby also that (1.13b) is not generally valid. Since under the conditions outlined in the above $\frac{\mathrm{d}}{\mathrm{d}t} = \frac{\mathrm{D}}{\mathrm{D}t}$, consequently, in the following we shall use the operator $\frac{\mathrm{d}}{\mathrm{d}t}$ to denote the substantial derivative. Sometimes the substantial derivative of a single quantity will be denoted by a point placed above the quantity instead, e.g. $\dot{A} \equiv \frac{\mathrm{d}A}{\mathrm{d}t}$.

Let us describe some concepts often used in field theory which follow from the given general formalism and refer to particular cases of practical importance. The point where

$$\boldsymbol{v} = 0 \tag{1.16}$$

is called the *stagnation point* of the field. If the velocity field is independent of time, we have

$$\boldsymbol{v} = \boldsymbol{v}(\boldsymbol{r}) \tag{1.17}$$

instead of (1.10), i.e. a velocity field is obtained which is constant in time. The motion characterized by such a velocity field is called *steady motion*. The concept of steady motion defined in (1.17) should not be confused with the concept of stationary motion, which will be interpreted in the following, and which gained a profound and many-sided meaning particularly in thermodynamics (cf. Chapter V). The concept of steady motion can be extended beyond the case of the velocity field to any field quantity A. Hence, if

$$A = A(\boldsymbol{r}) \tag{1.18}$$

we can say that the field of the quantity A is *steady*. Other, often used concepts can be interpreted by the particular cases of the relation (1.14). According to convention, any A field quantity is called:

a) *substantially constant* if

$$\dot{A} = 0, \tag{1.19}$$

i.e. if the field quantity A in the system of coordinates X_1, X_2, X_3 fixed to the "particle" is independent of time.

b) Any field quantity is called *locally constant* if

$$\frac{\partial A}{\partial t} = 0, \tag{1.20}$$

i.e. the field quantity A is constant at the point $\boldsymbol{r}$ of the system of coordinates x_1, x_2, x_3 fixed to the space: *the field there is stationary.*

c) Any field quantity is called *convective constant*, if

$$\boldsymbol{v} \cdot \nabla A = 0 \tag{1.21}$$

which is fulfilled with the exception of the trivial case $\boldsymbol{v} = 0$, either if $\boldsymbol{v}$ and ∇A are perpendicular to each other, or if $\nabla A = 0$. In the latter case it is generally said that the field of the quantity A is *homogeneous*.

5. Mass and the Material Equation of Continuity

In classical field theory we disregard molecular or atomic discontinuities and consider matter continuously distributed in the Euclidean space, and the mass is regarded as the most fundamental characteristic of it. The mass of each body can always be characterized by a positive real number which is the measure of inertia, gravity and quantity of matter included in the body in question. Considering mass as a measure of the quantity of matter has the mathematical meaning in the case of continua that:

the mass M is an absolutely continuous function of volume V [1].

This interpretation of mass includes also that the mass of a "body" with zero volume is zero too, further that the mass of a system is the sum of the masses of the constituent bodies. It should be noted that in thermodynamics quantities which similar to mass have additive properties are called *extensive* quantities.

Let us now interprete the most simple field quantity, i.e. the mass density ϱ. If the function $M = M(V)$ can be differentiated at the point $\boldsymbol{r}$ and if the limes

$$\frac{\mathrm{d}M}{\mathrm{d}V} = \lim_{\Delta V \to 0} \frac{\Delta M}{\Delta V} = \varrho(\boldsymbol{r}, t) \tag{1.22}$$

is finite the field quantity $\varrho = \varrho(\boldsymbol{r}, t)$ is the density of mass of the continuum at the point $\boldsymbol{r}$ and at the moment t. If the density function ϱ is known, the total mass of a body of finite volume V is given by the volume integral of the density in an arbitrary time t:

$$M(t) = \int_V \varrho(\boldsymbol{r}, t)\,\mathrm{d}V. \tag{1.23}$$

In classical field theory, "particle" means a mass element $\mathrm{d}M = \varrho\,\mathrm{d}V$ of the continuum and this is not identical with any of the particles occurring in corpuscular theories, (molecules, ions, atoms, elementary particles). From the viewpoint of corpuscular theories a $\mathrm{d}V$ volume element is a so-called "macro-differential", i.e. it is a region of an order of magnitude which is infinitesimal with respect to V, but at the same time it includes a sufficiently large number of corpuscles so that the individual motion of these can be disregarded. This idea is correct; however, our attention must be drawn to the fact that some views (which are unfortunately, adapted also by the authors of some excellent works), derived from it are incorrect, i.e. according to which the basic equations of field theories can be derived from the basic equations of corpuscular theories related to the mass points. This concept is wrong, not only because such derivations are in the majority of cases illusoric, but rather because they pretend that field theories (sometimes called phenomenological theories,though the two theories are not identical) are subordinate to corpuscular theories and that they play only a secondary role. Though this view is refuted by the widespread practical applicability of the field theories which is much more direct than that of corpuscular theories (as a matter of fact only in the case of macroscopical-size systems), the above mentioned and misleading views are, particularly in connection with thermodynamics, unfortunately rather widespread. It must be acknowledged that the relation between field theories relying upon the continuity axioms and the corpuscular theories is much more complicated, than it was thought to be. At present the fundamental difficulty lies in the contradiction between the field theory and corpuscular models formed for the description of nature. This is expressed by the fact that *the field is indefinitely divisible, whereas the corpuscle is not.*

Relations (1.22) and (1.23) are valid at any time. Hence, they are also valid when in a given initial time $t = t_0$ the density $\varrho_0 = \varrho_0(\boldsymbol{r}, t_0)$ is determined by the volume $\mathrm{d}V$ and by the mass $\mathrm{d}M$ present in it. If the conservation of mass is to be given for a certain $t - t_0$ time interval under a deformation $\mathrm{d}V \to \mathrm{d}V^*$ taking place owing to the motion of the material of the continuum, the knowledge of the function $\varrho = \varrho(\boldsymbol{r}, t)$ belonging to the time t is required. The conservation of mass for the deformation of the mass element in question is expressed according to (1.22) by condition

$$\mathrm{d}M = \varrho_0(\boldsymbol{r}, t_0)\,\mathrm{d}V = \varrho(\boldsymbol{r}, t)\,\mathrm{d}V^* = \text{const.} \tag{1.24}$$

for the deformation $\mathrm{d}V \to \mathrm{d}V^*$. Making use of the relation (1.5) the equation of conservation of mass (1.24) may be written in the form:

$$\varrho_0 = J\varrho. \tag{1.25}$$

This equation is called the *material* (Lagrangian) *equation of continuity.* The *spatial* (Eulerian) form of the continuity equation is derived in the following chapter. The equivalence of the two equations will also be confirmed there.

Finally, let us give the conservation of mass for the whole of a continuum, which always contains an identical mass during the motion causing a deformation $V \to V^*$. The global equation is obtained by the integration of (1.24) with respect to the corresponding volumes, hence

$$M = \int_V \varrho_0 \,\mathrm{d}V = \int_{V^*} \varrho \,\mathrm{d}V^* = \text{const.} \tag{1.26}$$

where M is the constant total mass of the material of the continuum during the motion.

6. Multi-component (superposed) Continua

From the viewpoints of chemistry, it is important to study continua which contain a number of different components in the chemical sense. Various diffusion phenomena, especially in the case of reacting components, require, the extension of the theory of single-component continua to the case of multi-component ones, in addition to a theory of chemical reactions.

In the sense of what has been said above, let us consider the general case, when the total mass M, continuously distributed in some finite region V of space, is the sum of the masses of chemical components, the number of which is K. In this case we have

$$M = \sum_{k=1}^{K} M_k, \tag{1.27}$$

where M_k, $(k = 1, 2, \ldots, K)$ is the mass of the k-th component which is assumed to be distributed continuously in the total volume according to the density: $\varrho_k = \varrho_k(\boldsymbol{r}, t)$. Speaking in the language of field theory, in our case K "particles" ($\mathrm{d}M_k$ mass elements) are present and carry out different individual deformations and motions in each internal point of the continuum.

Consequently, the continuous system must now be considered as the local superposition of K continuous media. According to this concept, the density of the k-th continuous medium (simply the k-th component) is defined by

$$\frac{\mathrm{d}M_k}{\mathrm{d}V} = \varrho_k(\boldsymbol{r}, t), \quad (k = 1, 2, \ldots, K) \tag{1.28}$$

from which it follows that the mass of the k-th component present in the total volume V is

$$M_k(t) = \int_V \varrho_k(\boldsymbol{r}, t)\, \mathrm{d}V, \qquad (k = 1, 2, \ldots, K). \tag{1.29}$$

The above-mentioned local superposition is expressed by equation

$$\varrho = \sum_{k=1}^{K} \varrho_k = \frac{1}{v} \tag{1.30}$$

valid for the densities, where $v = \varrho^{-1}$ is the specific volume of the compound continuum. Instead of the condition (1.30) we introduce the weight (or mass) fractions

$$c_k = \frac{\varrho_k}{\varrho}, \qquad (k = 1, 2, \ldots, K) \tag{1.31}$$

and often use the following normalized condition

$$\sum_{k=1}^{K} c_k = 1, \tag{1.32}$$

which is equivalent to (1.30).

It is evident that each component of the continua undergoes a deformation in a different way. Accordingly, the mass elements $\mathrm{d}M_k$ move individually and have different individual velocities. These motions are described, similarly to (1.6), by

$$\boldsymbol{r} = \boldsymbol{r}(\boldsymbol{R}_k, t) \tag{1.33}$$

where $\boldsymbol{R}_k$ is the "particle" of the k-th component medium in the sense of the material description. The velocity is defined similarly to (1.8), hence

$$\boldsymbol{v}_k \equiv \left(\frac{\partial \boldsymbol{r}}{\partial t}\right)_{\boldsymbol{R}_k}, \qquad (k = 1, 2, \ldots, K) \tag{1.34}$$

is the individual velocity of the "particle" $\boldsymbol{R}_k$ of the k-th component. The individual velocity fields corresponding to the spatial description are obtained, if from the set of functions

$$\boldsymbol{v}_k = \boldsymbol{v}_k(\boldsymbol{R}_k, t), \qquad (k = 1, 2, \ldots, K) \tag{1.35}$$

$\boldsymbol{R}_k$ is eliminated with the inverse transformation $\boldsymbol{R}_k = \boldsymbol{R}_k(\boldsymbol{r}, t)$ of (1.33):

$$\boldsymbol{v}_k = \boldsymbol{v}_k(\boldsymbol{r}, t), \qquad (k = 1, 2, \ldots, K). \tag{1.36}$$

It can be seen that these velocity fields of the component media are not independent of each other since, owing to the condition (1.30) (or (1.32)) of local superposition, they determine the barycentric (mean) velocity $\boldsymbol{v}$ of the resultant continuum. The barycentric velocity $\boldsymbol{v}$ of the "particle" $\boldsymbol{R}$ of the superposed continuum is determined by the individual velocities $\boldsymbol{v}_1, \boldsymbol{v}_2, \ldots, \boldsymbol{v}_k$ of the "particles" $\boldsymbol{R}_1, \boldsymbol{R}_2, \ldots, \boldsymbol{R}_k$ of the component continua, by the following requirement. The sum of the individual and local mass current densities

$$\boldsymbol{J}_k^0(\boldsymbol{r}, t) = \varrho_k(\boldsymbol{r}, t)\,\boldsymbol{v}_k(\boldsymbol{r}, t), \qquad (k = 1, 2, \ldots, K) \tag{1.37}$$

must be equal to the local mass current density of the resultant continuum

$$\boldsymbol{J}^0(\boldsymbol{r}, t) = \varrho(\boldsymbol{r}, t)\,\boldsymbol{v}(\boldsymbol{r}, t) \tag{1.38}$$

hence, the relation

$$\varrho\boldsymbol{v} = \sum_{k=1}^{K} \varrho_k \boldsymbol{v}_k, \tag{1.39a}$$

or

$$\boldsymbol{v} = \sum_{k=1}^{K} c_k \boldsymbol{v}_k \tag{1.39b}$$

respectively, must be fulfilled. These relations can be also considered as the definition of the barycentric velocity $\boldsymbol{v}$ in case of multi-component continua.

From the point of view of corpuscular theories (for instance, Boltzmann's kinetic theory) the individual $\boldsymbol{v}_k$ velocity is the velocity of the mass element $\mathrm{d}M_k$, which at the moment t is at the place $\boldsymbol{r}$ of space, and which contain yet a very large number of identical corpuscles (atoms, molecules, ions, etc.). Consequently, the velocities $\boldsymbol{v}_k$ are related to the corresponding corpuscle velocities $\mathfrak{v}_k$ according to relation

$$\boldsymbol{v}_k(\boldsymbol{r}, t) \leftrightarrow \langle \mathfrak{v}_k(\boldsymbol{r}, t)\rangle = \int \mathfrak{v}_k f_k(\boldsymbol{r}, \mathfrak{v}_k, t)\, \mathrm{d}^3\mathfrak{v}_k \tag{1.40}$$

where f_k is a velocity distribution function of the k-th kind of corpuscle, for which the Boltzmann transport equation is satisfied [6]. Taking into account the corpuscular theories, it can be said, that the velocity fields $\boldsymbol{v}_k$ can be represented by the mean values of the velocity $\mathfrak{v}_k$ of the corpuscles forming the mass element $\mathrm{d}M_k$. It follows both from the phenomenological considerations and from the molecular picture (which is perhaps more transparent but unnecessary for the description of the theory of fields) that the individual $\boldsymbol{v}_k$ velocities of the different component media result macroscopically in the diffusion phenomena.

Let us define the diffusion velocity and the current density of the k-th component. Since according to (1.39) the $\boldsymbol{v}$ barycentric velocity of the "particle" $\mathrm{d}M$ is determined by the individual velocity $\boldsymbol{v}_k$ of the "particles" $\mathrm{d}M_k$, it is preferable to define the diffusion velocities $\boldsymbol{w}_k$ and the current densities $\boldsymbol{J}_k \equiv \varrho_k \boldsymbol{w}_k$ with respect to the barycentric velocity. Hence, by definition

$$\boldsymbol{w}_k \equiv \boldsymbol{v}_k - \boldsymbol{v}, \qquad (k = 1, 2, \ldots, K) \tag{1.41}$$

and

$$\boldsymbol{J}_k \equiv \varrho_k \boldsymbol{w}_k = \varrho_k(\boldsymbol{v}_k - \boldsymbol{v}), \qquad (k = 1, 2, \ldots, K) \tag{1.42}$$

are the diffusion velocity and current density, respectively, referred to the local barycentric velocity. According to the given definitions the individual motion of $\boldsymbol{v}_k$ the velocity of the k-th component consists of two parts: of the motion of the barycentric velocity $\boldsymbol{v}$ and of the diffusion velocity $\boldsymbol{w}_k$ relative to the former. It also follows from (1.39) and (1.42) that the diffusion current densities are not all independent, because the number of the independent current densities is reduced to $K-1$ by the relation [obtained by the summing of (1.42)]:

$$\sum_{k=1}^{K} \boldsymbol{J}_k = \sum_{k=1}^{K} \varrho_k \boldsymbol{w}_k = \sum_{k=1}^{K} \varrho_k (\boldsymbol{v}_k - \boldsymbol{v}) = 0 . \tag{1.43}$$

This relation represents the existence of a local constraint for the diffusion current densities, which is a consequence of relations (1.39).

Finally, with respect to a scalar field quantity A a relation similar to (1.14) is given, which refers to the individual motion of the k-th component medium. If, similarly to (1.14), the substantial time derivative referred to the k-th medium moving with a velocity $\boldsymbol{v}_k$ is introduced,

$$\frac{\mathrm{d}^{(k)} A}{\mathrm{d}t} = \left(\frac{\partial A}{\partial t}\right)_r + \boldsymbol{v}_k \cdot \nabla A \tag{1.44}$$

is the required relation. Here $\frac{\mathrm{d}^{(k)}}{\mathrm{d}t}$ is the substantial differential operator referring to the k-th medium, whereas $\boldsymbol{v}_k \cdot \nabla A$ measures the variation of the field quantity A, which results from the convection of the k-th continuum with a velocity $\boldsymbol{v}_k$. The relation

$$\frac{\mathrm{d}^{(k)} A}{\mathrm{d}t} - \frac{\mathrm{d}A}{\mathrm{d}t} = \boldsymbol{w}_k \cdot \nabla A \tag{1.45}$$

obtained as the difference of (1.14) and (1.44) will be employed later. It must be emphasized that in the relations (1.11) to (1.21) and similarly in relations (1.44) and (1.45) it has been assumed that the field quantity A is a scalar in order to avoid the cumbersome tensorial description. If later on some of the given relations will be applied to a vector or to any field quantity of a higher tensorial order, it must always be applied to the scalars stipulated by the order of the tensor in question. The reader is asked to bear this in mind.

II. Balance Equations

In this chapter, general balance equations of fundamental importance in field theories are given in the local as well as in the substantial form. First the relations existing between them are described and then the detailed treatment of balance equations necessary for the development of thermodynamics in terms of concepts of field theories are treated. Mass, charge, impulse and angular momentum balances are discussed in detail, followed by a description of the different energy balances for multi-component systems. By these balance equations it is possible to determine the entropy balance (Chapter III), which plays a central role in thermodynamics and lends itself for the treatment of multi-component and reacting hydro-thermodynamic systems which are of particular importance in chemical industry, plasma-physics, biology, etc. Beyond this we endeavour to determine generalized forms of the balance equations which should be applicable also for system-models, whose theoretical thermodynamic treatment is the task of the science of today. In this respect, first of all, the so-called thermo-mechanical theory of plastic materials and rheological systems as well as the thermo-electrodynamics of dielectrics may be mentioned.

1. General Balance Equations

Let A be some arbitrary extensive quantity, and a its specific value referred to unit mass. If the field quantity A is distributed in the material of a continuum whose volume and density are V and ϱ, the total variation of the quantity

$$A = \int_V \varrho a \,\mathrm{d}V \tag{2.1}$$

in the course of time

$$\dot{A} = \frac{\mathrm{d}}{\mathrm{d}t} \int_V \varrho\, a \,\mathrm{d}V \tag{2.2}$$

can arise, in general, for two reasons (Fig. 2).

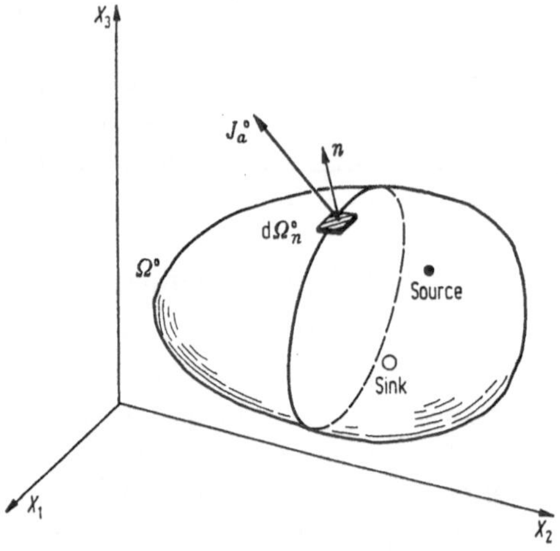

Fig. 2

1. Owing to the influx or outflux of A through the Ω boundary-surface of the volume V.

2. Owing to the decrease or increase (production) of the quantity of A inside the volume V, which is the consequence of sinks or sources for the quantity A present in the internal points of the continuum.

The determination of the general balance equations is effected on the basis of these points. Two different types of balance equations are obtained; depending on whether the spatial or the material description is chosen as a primary one: the *local* and the *substantial* forms of the balance equations.

a) Local Balances. The general form of the local balance equations based upon the spatial description is obtained by the assumption that the volume with respect to which the variation of a quantity A must be expressed is at rest relative to the external (Eulerian) coordinate system. In this case, instead of (2.2), we can write

$$\frac{\mathrm{d}}{\mathrm{d}t}\int_{V^0} \varrho a \,\mathrm{d}V^0 = \int_{V^0} \frac{\partial \varrho a}{\partial t}\,\mathrm{d}V^0 \tag{2.3}$$

where the integrations should be carried out over the volume elements $\mathrm{d}V^0 = \mathrm{d}x_1\,\mathrm{d}x_2\,\mathrm{d}x_3$ of constant position in the x_1, x_2, x_3 coordinate system. If in some medium the density of which is ϱ the transport of a field quantity A occurs, the intensity of this transport is characterizable by a $\boldsymbol{J}_a^0$ vector

$$\boldsymbol{J}_a^0 = \varrho a \boldsymbol{v}_a, \tag{2.4}$$

which is the local current density of the field quantity A, and where $\boldsymbol{v}_a$ is the velocity of the transport of A. The current density $\boldsymbol{J}_a^0$ is the quantity of A crossing a plane of unit area in unit time when the plane is fixed relative to the external coordinate system and is oriented normal to the local current density $\boldsymbol{J}_a^0$. Denoting the density of the internal source of A by σ_a, and on the basis of points 1. and 2., the relation

$$\int_{V^0} \frac{\partial \varrho a}{\partial t}\,\mathrm{d}V^0 = -\oint_{\Omega^0} \boldsymbol{J}_a^0 \cdot \mathrm{d}\boldsymbol{\Omega}^0 + \int_{V^0} \sigma_a \,\mathrm{d}V^0 \tag{2.5}$$

defines the integrated (global) form of the local type balance equation. Here $\mathrm{d}\boldsymbol{\Omega}^0 = \mathrm{d}\Omega^0 \boldsymbol{n}$ is a vectorial surface element, whose magnitude is $\mathrm{d}\Omega^0$ and its direction is determined by the outward normal $\boldsymbol{n}$ (Fig. 2).

Balance equation (2.5) cannot be confirmed directly for any quantity. Its consequences can be compared to experience at least in some cases. In principle, (2.5) and any other balance equation is nothing but a simultaneous definition for the quantities $\int_{V^0} \frac{\partial \varrho a}{\partial t}\,\mathrm{d}V^0$, $\oint_{\Omega^0} \boldsymbol{J}_a^0 \cdot \mathrm{d}\boldsymbol{\Omega}^0$

and $\int\limits_{V^0} \sigma_a \, \mathrm{d}V^0$ and for their relation. Its significance lies in the fact that given the other two quantities any of the quantities can be determined. In general, in the case of a given variation of A the quantities $\boldsymbol{J}_a^0$ and σ_a can be appropriately defined, and thus it may be said that all A quantities can be balanced in different ways, by definition.

The above-mentioned is of particular importance from the viewpoints of the formulation of the so-called conservation laws of physics. The question whether a quantity for a given continuum might be considered as conservative, can be answered only in the case of an unambiguous determination (separation) of $\boldsymbol{J}_a^0$ and σ_a. In such cases, particular attention must be given to points 1. and 2., which is not a simple task in the case of quantities similar to such abstract concepts as, for example, internal energy and entropy. Hence, selected attention must be directed to the questions related to the defining character of the balance equations especially when establishing internal energy and entropy balances. The fact that the formulation of the balance equations of these quantities is sometimes different in the literature, and is even carried out in an incorrect form by chance, is due to the unsatisfactory analysis of the conditions of the model of the systems examined.

Let us determine the differential form of balance equation (2.5), i.e. the form valid in each internal point of the continuum. Transforming the surface integral on the right-hand side of (2.5) with the Gauss' divergence theorem to volume integral

$$\oint\limits_{\Omega^0} \boldsymbol{J}_a^0 \cdot \mathrm{d}\boldsymbol{\Omega}^0 = \int\limits_{V^0} \nabla \cdot \boldsymbol{J}_a^0 \, \mathrm{d}V^0 \tag{2.6}$$

the balance (2.5) can be given in a zero reduced form:

$$\int\limits_{V^0} \left(\frac{\partial \varrho a}{\partial t} + \nabla \cdot \boldsymbol{J}_a^0 - \sigma_a \right) \mathrm{d}V^0 = 0 . \tag{2.7}$$

Since this relation must be valid for an arbitrary volume which is at rest with respect to the x_1, x_2, x_3 coordinate system, the differential equation

$$\frac{\partial \varrho a}{\partial t} + \nabla \cdot \boldsymbol{J}_a^0 = \sigma_a \tag{2.8}$$

is obtained, which is called the *local form of the differential balance equation* referred to the field quantity A.

With respect to the differential equation (2.8) the foregoings concerning the defining character of the integrated form (2.5) leads to the result that in the case of a given $\frac{\partial \varrho a}{\partial t}$ the source density σ_a can uniquely be separated from the current density $\boldsymbol{J}_a^0$ if and only if the divergence of the latter i.e. $\nabla \cdot \boldsymbol{J}_a^0$ is known. Hence, in the case of a given $\frac{\partial \varrho a}{\partial t}$ and with the knowledge of $\nabla \cdot \boldsymbol{J}_a^0$, the σ_a can uniquely be defined and the problem of conservation of any field quantity A can be unambigously determined.

If the source density σ_a of some A vanishes locally: $\sigma_a = 0$, it can be stated that A is conserved locally. Otherwise if $\sigma_a > 0$, there is the question of the local production of A, whereas if $\sigma_a < 0$ we speak of the local absorption of A. The cases in question have also a sense for the global system if the source density σ_a vanishes, is positive or negative in the whole volume V.

Let us apply the above-mentioned in connection with the balance expressing the mass conservation of the continuum. If $A \equiv M$ is the total mass of the system, then $a \equiv 1$ is the specific mass and we obtain from (2.4) the local mass current density

$$\boldsymbol{J}^0 = \varrho \boldsymbol{v}, \tag{2.9}$$

which was already defined in a different way in (1.38). On the other hand, for this particular case from (2.8) we have

$$\frac{\partial \varrho}{\partial t} + \nabla \cdot \varrho \boldsymbol{v} = 0 \tag{2.10a}$$

or

$$\frac{\partial \varrho}{\partial t} + \nabla \cdot \boldsymbol{J}^0 = 0, \tag{2.10b}$$

by which the conservation of mass is expressed for an arbitrary internal point of the continuum. The physical content of this local and source-free mass balance is that *the production (or decrease) of mass present in the unit volume of the continuum is equal to the influx (or efflux resp.) of the mass quantity with the current density $\boldsymbol{J}^0 = \varrho \boldsymbol{v}$*. The differential equations (2.10) are often called the spatial (or local) equation of continuity. The more profound sense of the denomination will be enlightened in the following point.

b) Substantial Balances. The general forms of the substantial balance equations based on the material description are obtained, if the selected volume element moves together with the medium, which in turn moves with a velocity $\boldsymbol{v}$ in the system of coordinates x_1, x_2, x_3 fixed to the space. The co-motion of the mass element $\mathrm{d}M = \varrho\,\mathrm{d}V$ with the volume element $\mathrm{d}V$ at a velocity $\boldsymbol{v}$ means that during the motion $\mathrm{d}V$ always contains a constant mass $\mathrm{d}M$. Strictly speaking it is always a "particle" of the same mass $\mathrm{d}M$ which is present during motion in the volume element $\mathrm{d}V$. This physical picture can be expressed mathematically by the condition that, when expressing the local mass current $\boldsymbol{J}^0$ by the substantial mass current $\boldsymbol{J}$, the latter must be identical to zero, i.e. the condition

$$\boldsymbol{J} = \boldsymbol{J}^0 - \varrho \boldsymbol{v} \equiv 0 \tag{2.11}$$

must be fulfilled. Of course, this condition does not mean that in case of the material description, the substantial current densities of any field quantities A would vanish. It is even shown by (2.4), (2.9) and (2.11) that the substantial current density of an arbitrary quantity A must be defined by the relation

$$\boldsymbol{J}_a = \boldsymbol{J}_a^0 - \varrho a \boldsymbol{v} = \varrho a(\boldsymbol{v}_a - \boldsymbol{v}). \tag{2.12}$$

The correctness of this definition is shown by the fact that in case of $a \equiv 1$, when $\boldsymbol{J}_a$ is identical to the mass current density, (2.12) becomes zero in accordance with (2.11).

In case of the material description, the $\mathrm{d}V$ volume element is always filled by the same mass element $\mathrm{d}M = \varrho\,\mathrm{d}V$. This means that in the course of motion $\mathrm{d}M$ is a quantity constant in time, and therefore, instead of (2.3) we can write

$$\frac{\mathrm{d}}{\mathrm{d}t}\int_V \varrho a\,\mathrm{d}V = \int_V \varrho\dot{a}\,\mathrm{d}V, \tag{2.13}$$

because in this case the substantial differentiation with respect to time effects the quantity a only. In this expression the integration has to be extended to the volume V moving with the continuum. Considering the contents of points 1. and 2., equations (2.12) and (2.13) yield, the following balance equation

$$\int_V \varrho\dot{a}\,\mathrm{d}V = -\oint_\Omega \boldsymbol{J}_a \cdot \mathrm{d}\boldsymbol{\Omega} + \int_V \sigma_a\,\mathrm{d}V, \tag{2.14}$$

which is called, the *integrated (global) form of the substantial balance equation* because of the employed material description. From this, by application of the Gauss theorem, we have

$$\varrho\dot{a} + \nabla \cdot \boldsymbol{J}_a = \sigma_a \tag{2.15}$$

which is the *differential form of the substantial balance.*

Let us derive a relation by which the local and substantial variation of an arbitrary specific field quantity a is connected, i.e. by which a bridge is formed between the spatial and material description. Applying the operator equation (1.15) for ϱ, we have

$$\dot{\varrho} = \frac{\partial\varrho}{\partial t} + \boldsymbol{v} \cdot \nabla\varrho, \tag{2.16}$$

by which from (2.10) the substantial and source-free balance equation

$$\dot{\varrho} + \varrho\nabla\cdot\boldsymbol{v} = 0 \tag{2.17}$$

is obtained. This expression describes, just as (2.10), the conservation of mass of the continuum. Repeatedly operating with (1.15) for ϱa, we get

$$\frac{\mathrm{d}\varrho a}{\mathrm{d}t} = \dot{\varrho}a + \varrho\dot{a} = \frac{\partial\varrho a}{\partial t} + \boldsymbol{v} \cdot \nabla\varrho a, \tag{2.18}$$

from which by eliminating $\dot{\varrho}$ with (2.17) the relation

$$\varrho \dot{a} = \frac{\partial \varrho a}{\partial t} + \varrho a \nabla \cdot \boldsymbol{v} + \boldsymbol{v} \cdot \nabla \varrho a \tag{2.19}$$

is obtained. Transforming the right-hand side of this equation with the identity

$$\nabla \cdot \varrho a \boldsymbol{v} = \varrho a \nabla \cdot \boldsymbol{v} + \boldsymbol{v} \cdot \nabla \varrho a \tag{2.20}$$

we get, so our aim,

$$\varrho \dot{a} = \frac{\partial \varrho a}{\partial t} + \nabla \cdot \varrho a \boldsymbol{v}, \tag{2.21}$$

the relation between the substantial and local variation of a. We emphasize that in (2.21) a can be any scalar of some specific field quantity. Hence a may be a scalar or a component of a vector or a second order tensor, etc. If, as a special case, the specific mass is $a \equiv 1$, (2.21) is reduced to the local mass balance (2.10).

The relation (2.21) will be often used later, but a fundamental question will be clarified by it now. Namely, whether the local (2.8) and substantial balance equations (2.15) are equivalent to one another. The proof is needed, because when deriving (2.15) it was assumed that the source density σ_a is identical to that given in the local equation (2.8). This requires verification which can immediately be done by (2.21). Adding the terms of zero sum $\nabla \cdot \varrho a \boldsymbol{v} - \nabla \cdot \varrho a \boldsymbol{v}$ to (2.8) we can write

$$\left[\frac{\partial \varrho a}{\partial t} + \nabla \cdot \varrho a \boldsymbol{v}\right] + \nabla \cdot \boldsymbol{J}_a^0 - \nabla \cdot \varrho a \boldsymbol{v} = \sigma_a, \tag{2.22}$$

which is from (2.21) and (2.12) just the substantial balance equation (2.15). Hereby, the equivalence of (2.8) and (2.15) has been confirmed by the result that the respective source terms are identical.

The equivalence of the local (2.10) and substantial (2.17) balance equations expressing the conservation of mass also follows from the foregoing. These equations are often called spatial (Eulerian) continuity equations. In this connection, we mention that for instance, the equivalence of (2.17) with the material (Lagrangian) equation of continuity (1.25) can be proved. Let us consider the cofactor $\mathrm{D}_{\alpha\beta}$ belonging to the Jacobian (1.3) and by which the Jacobian can be expressed as follows:

$$\frac{\partial x_\beta}{\partial X_\alpha} \mathrm{D}_{\alpha\gamma} = J\, \delta_{\beta\gamma}, \qquad \delta_{\beta\gamma} = \begin{cases} 0 & \beta \neq \gamma \\ 1 & \beta = \gamma \end{cases} \tag{2.23}$$

where $\delta_{\beta\gamma}$ is the Kronecker symbol.[1] By using some elementary rules of determinant theory and (2.23), we get for the substantial time deri-

[1] Now and sometimes later as well (but not in all cases) we use the convention that a repeated index is automatically summed from 1 to 3, according to the standard notation of tensor analysis.

vative of the Jacobian

$$\frac{\mathrm{d}J}{\mathrm{d}t} = \frac{\mathrm{d}}{\mathrm{d}t}\left(\frac{\partial x_\beta}{\partial X_\alpha}\right) \mathrm{D}_{\alpha\beta} = \frac{\partial v_\beta}{\partial X_\alpha} \mathrm{D}_{\alpha\beta} = \frac{\partial v_\beta}{\partial x_\gamma}\frac{\partial x_\gamma}{\partial X_\alpha} \mathrm{D}_{\alpha\beta} = \frac{\partial v_\beta}{\partial x_\beta} J, \tag{2.24}$$

which can also be written in a concise vectorial form:

$$\dot{J} = J \boldsymbol{\nabla} \cdot \boldsymbol{v}. \tag{2.25}$$

By this elegant relation originally due to Euler, the equivalence of (2.17) with the material continuity equation (1.25) can be demonstrated immediately. By multiplying (2.17) with the Jacobian J and making use of the relation (2.25), equation

$$\dot{\varrho} J + \varrho \dot{J} = 0$$

is obtained, which can also be written in the forms

$$\frac{\mathrm{d}\varrho J}{\mathrm{d}t} = 0 \tag{2.26a}$$

or

$$\varrho J = \varrho_0 \tag{2.26b}$$

if ϱ_0 is the density distribution belonging to the initial time. Hereby, the equivalence of (2.17) (and also (2.10)) with the material continuity equation has been proved. Consequently, equations (2.10) and (2.17) can be called the *spatial equations of continuity.*

It should be noted that this denomination is often used by some authors for balance equations given for arbitrary field quantities as well. In this sense we speak, apart from continuity equations valid for mass, about continuity equations referring to charge, impulse, angular momentum, different species of energies and entropy. In the following we shall use the denomination "balance equation" in spite of the fact that the very widespread denomination *continuity equation* instead of *balance equation* is in the sense of the foregoing correct. Moreover it emphasizes an important fact, viz. that according to the field theory the continuity axiom required for mass distribution is also extended to different properties of the continuously distributed material.

As regards the conservation of an arbitrary field quantity A our previous statements have to be completed. In general, balance equations (2.8) and (2.15) represent conservation laws, hence, a field quantity A is called conservative, if

$$\frac{\partial \varrho a}{\partial t} + \boldsymbol{\nabla} \cdot \boldsymbol{J}_a^0 = 0 \tag{2.27}$$

or the equivalent

$$\varrho \dot{a} + \boldsymbol{\nabla} \cdot \boldsymbol{J}_a = 0 \tag{2.28}$$

equation is valid in every point of the continuum. It is self-evident that an infinite number of current density fields $\boldsymbol{J}_a^0(\boldsymbol{r}, t)$ and $\boldsymbol{J}_a(\boldsymbol{r}, t)$ can be given in such a way that the conservation laws (2.27) and (2.28) are fulfilled.

In cases, where the question of the conservation of some quantity A, is raised globally i.e. concerning the system as a whole, the isolation conditions determining the system-model must be defined exactly. On the other hand, for the sake of a clear formulation of the conservation laws, it is advisable to divide the source densities σ_a into an "internal" σ_a^{i} and an "external" σ_a^{e} production density according to the sum:

$$\sigma_a = \sigma_a^{\mathrm{i}} + \sigma_a^{\mathrm{e}}. \tag{2.29}$$

Quantities of the total source density σ_a which are determined by the inhomogeneities existing locally in the interior of the system, belong to the "internal" source density σ_a^{i}. Such local internal inhomogeneities may arise as a result of an inhomogeneous distribution of velocity, temperature, chemical potential, etc. Consequently, these internal inhomogeneities can always be measured by gradients (velocity gradient, temperature gradient, chemical potential gradient etc.) which lead to the "internal" source density σ_a^{i}. The gradients representing the internal inhomogeneities in question and the current densities caused by them are the factors, which determine the "internal" source density σ_a^{i}. Since, the "internal" source density σ_a^{i} of arbitrary quantities A is a function of the given inhomogeneity in the internal of the system, thus it is evident that irreversible processes are always related to "internal" source densities. The foregoings are particularly important for the understanding of the balances of entropy.

The situation is quite different for the interpretation of the so-called "external" source density σ_a^{e}. The "external" source density of a field quantity A is always a result of the long-range action of external forces acting on the system, but, of course, also in the interior of the system. However, "external" source densities of the impulse and energy of the external fields (gravitational, electromagnetic etc.) acting on the material of the system are always existing. By taking this into account or by chosing our system-model appropriately so that the external fields are considered to belong to the system, the conservation of the field quantity A in question can be considered in the case of non-vanishing "external" source densities σ_a^{e}. This is the situation, for instance, in the case of impulse balance expressed by the equation of the motion of the continuum. Nevertheless, let us note that a system-model subjected to a scientific analysis must in every case be fixed by exact and well-defined conditions, furthermore, that this is particularly necessary in thermodynamics, because in this discipline one works with an abundance of different system-models.

2. Balances of Mass

The local (2.10) and substantial (2.17) mass balances are quite general and express the conservation of mass of the continuous system independent of whether the system contains a single material or includes the local superposition of a media with many different chemical components. In the case of such mixtures, the mass balances (2.10) and (2.17) require a further specification. Here mass balances are stated also for every single component of the mixtures by which we can describe the diffusion phenomena and occasionally the chemical reactions as well. At the same time they have to be compatible with the balance equation expressing the conservation of the total mass.

Such a generalization of mass balances, viz. that diffusion phenomena and chemical reactions can be dealt with, is based upon the following two conditions.

a) It is assumed that the total mass M of the continuum is composed, in accordance with relation (1.27), of the sum of K number of chemical components of a mass M_k.

b) It is assumed that between the K components R chemical reactions take place in the form of a local internal transformation in every point of the system.

As regards the second condition, it should be noted that in agreement with the spirit of the field theories, chemical reactions occuring in a multi-component continuum must be considered as transformations taking place between the internal degrees of freedom of molecules at any point of the continuum.[1]

Now let us give the mass balance valid for individual components of a continuous system of K components, which are often called *component balances*. The local mass balance of the k-th component can immediately be written as a special case of the general local balance (2.8). From (2.8) with the choice of

$$a \equiv \frac{\varrho_k}{\varrho} = c_k, \quad \boldsymbol{J}_a^0 \equiv \boldsymbol{J}_k^0, \quad \sigma_a \equiv m_k \tag{2.30}$$

and with the local component current densities $\boldsymbol{J}_k^0$ defined in (1.37), the local component balances

$$\frac{\partial \varrho_k}{\partial t} + \nabla \cdot \boldsymbol{J}_k^0 = m_k, \qquad (k = 1, 2, \ldots, K) \tag{2.31}$$

are obtained. As will be seen later, the source terms m_k occuring here are in a simple relation to the chemical reaction rates and give the mass

[1] The fundamentals of such thermodynamic theory based upon the redistribution of the internal degrees of freedom have been developed by PRIGOGINE and MAZUR [7]; see also [3].

variation due to chemical transformations of the k-th component per unit volume and unit time. Of course, for all non-reacting (inert) components $m_k \equiv 0$.

Summing the component balances (2.31) over all the components, and taking into consideration relations (1.30) and (1.39) valid in the case of the superposed continuum, we have

$$\sum_{k=1}^{K}\left(\frac{\partial \varrho_k}{\partial t} + \nabla \cdot \boldsymbol{J}_k^0\right) = \sum_{k=1}^{K} m_k = 0, \tag{2.32}$$

which means the conservation of the total mass. This equation is in the case of multi-component continua equivalent to (2.10).

The substantial forms of the component balances can easily be determined. Eliminating the local derivative $\frac{\partial \varrho_k}{\partial t}$ from (2.31) with the operator equation (1.15), and introducing simultaneously the diffusional current densities $\boldsymbol{J}_k$ defined in (1.42), we have

$$\dot{\varrho}_k + \varrho_k \nabla \cdot \boldsymbol{v} + \nabla \cdot \boldsymbol{J}_k = m_k, \qquad (k = 1, 2, \ldots, K). \tag{2.33}$$

Summing now over all the components and taking relations (1.30) and (1.43) into consideration we come back to the balances equation of the total mass (2.17). This should have been expected on the basis of the conservation of the total mass. It is usual to write the component balances (2.33) in terms of mass fractions

$$\varrho \dot{c}_k + \nabla \cdot \boldsymbol{J}_k = m_k, \qquad (k = 1, 2, \ldots, K) \tag{2.34}$$

which is a form more concise than that of (2.33). This equation can be obtained from (2.33) with (1.31) and (2.17). We note that the general balance (2.15) leads directly to (2.34) by chosing $a \equiv c_k$.

If no chemical transformations take place between the components, or we disregard their description, all source terms in the component balances vanish identically, i.e. $m_k \equiv 0$ for every k. In such cases, each mass M_k of the components is an individual conservative quantity. Whereas, if chemical reactions take place between the components, the m_k source densities of the components do not vanish, at least not all of them.

Let us consider the case where in every internal point of the continuum R chemical reactions take place according to the stoichiometric equations

$$\sum_{k=1}^{r} \nu_{kj}^{c} \mathsf{M}_k \rightleftarrows \sum_{k=r+1}^{K} \nu_{kj}^{c} \mathsf{M}_k, \qquad (j = 1, 2, \ldots, R). \tag{2.35}$$

Here M_k is the molecular mass of the k-th component, whereas ν_{kj}^{c} is the stoichiometric number of the k-th component in the j-th reaction. By convention these are taken positive, if the component appears in the second term ($k = r + 1, r + 2, \ldots, K$) of the reaction equation (i.e. when it is a "product" of the reaction j), and negative when it appears in the first term ($k = 1, 2, \ldots, r$) (i.e. when it is a "reactant" in the reaction j). Thus the conservation laws of mass for chemical transformations described by the reaction (2.35) are expressed by

$$\sum_{k=1}^{K} \nu_{kj}^{c} \mathbf{M}_k = 0, \qquad (j = 1, 2, \ldots, R) \tag{2.36}$$

independent of the fact, whether the chemical reactions represented by (2.35) are independent of each other or not. The latter problem can be decided by the examination of the rank of the so-called stoichiometric matrix, which can be formed from the stoichiometric numbers ν_{kj}^{c}, which will not be dealt with here. Our task is only to express the source terms m_k in terms of the chemical reaction rates, and thus to take into account the transformations of matter due to the chemical reactions in the field equations.

First of all we introduce, instead of the stoichiometric coefficients used in chemistry, the following new coefficients

$$\nu_{kj} = \frac{\nu_{kj}^{c} \mathbf{M}_k}{\sum\limits_{k=r+1}^{K} \nu_{kj}^{c} \mathbf{M}_k}, \qquad (k = 1, 2, \ldots, K; \quad j = 1, 2, \ldots, R) \tag{2.37}$$

by which the equations (2.36) can be described in the following form [8]:

$$\sum_{k=1}^{K} \nu_{kj} = 0, \qquad (j = 1, 2, \ldots, R). \tag{2.38}$$

The coefficients ν_{kj} can always be determined with the knowledge of the ordinary chemical stoichiometric numbers ν_{kj}^{c}. On the other hand, they are more suitable for the definition of the chemical reaction rates interpreted locally in case of continua than the ordinary stoichiometric numbers.

Let us define the reaction rate of the j-th reaction referred to an internal point of the continuum. According to de Donder, if $\partial_j M_k$ is the mass variation of the k-th component in the j-th reaction, the relation between the mass variation $\partial_j M_k$ and the $\partial \xi_j$ variation of the reaction coordinate ξ_j in the j-th reaction is in the case of a homogeneous system of the mass M described by the following equation [8, 9]:

$$\partial_j M_k = M \nu_{kj} \, \partial \xi_j, \qquad (k = 1, 2, \ldots, K; \quad j = 1, 2, \ldots, R). \tag{2.39}$$

This relation must be generalized for inhomogeneous continua. Let ΔV be a volume element of the inhomogeneous continuum, to which

$\Delta M = \varrho \Delta V$, i.e. a mass quantity belongs. If in the same sense ΔM_k is the mass element of the k-th component, the relation

$$\partial_j \Delta M_k = \Delta M \nu_{kj} \, \partial \xi_j, \qquad (k = 1, 2, \ldots, K; \quad j = 1, 2, \ldots, R) \tag{2.40}$$

should be used instead of (2.39) with respect to the selected volume element ΔV.

By this equation, the rate of the j-th reaction can already be interpreted per unit volume as

$$J_j = \frac{\partial \xi_j}{\partial t} = \frac{1}{\nu_{kj} \Delta V} \frac{\partial_j \Delta M_k}{\partial t} = \frac{1}{\nu_{kj}} \frac{\partial_j}{\partial t} \left(\frac{\Delta M_k}{\Delta V} \right), \tag{2.41}$$

which leads, in the limiting case $\Delta V \to 0$ and owing to (1.28), by definition to

$$J_j = \frac{1}{\nu_{kj}} \frac{\partial_j \varrho_k}{\partial t}, \qquad (j = 1, 2, \ldots, R) \tag{2.42}$$

giving the local rate of the j-th reaction. For the local mass variation of the k-th component in all R reactions, we have

$$\frac{\partial \varrho_k}{\partial t} = \sum_{j=1}^{R} \frac{\partial_j \varrho_k}{\partial t} = \sum_{j=1}^{R} \nu_{kj} J_j, \qquad (k = 1, 2, \ldots, K) \tag{2.43}$$

if the density variation of the components is in some arbitrary point of space solely a consequence of the chemical transformations. However, the source densities m_k per unit volume and unit time are determined by

$$m_k = \sum_{j=1}^{R} \nu_{kj} J_j, \qquad (k = 1, 2, \ldots, K) \tag{2.44}$$

in every case.

The general validity of (2.44) allows that the mass balances be given also in cases, when besides the diffusion of K components R chemical reactions occur in the system in the form of local internal transformations. In such cases, the mass balances (2.31) and (2.34) must be used in the form

$$\frac{\partial \varrho_k}{\partial t} + \nabla \cdot \boldsymbol{J}_k^0 = \sum_{j=1}^{R} \nu_{kj} J_j \qquad (k = 1, 2, \ldots, K) \tag{2.45}$$

and

$$\varrho \dot{c}_k + \nabla \cdot \boldsymbol{J}_k = \sum_{j=1}^{R} \nu_{kj} J_j, \qquad (k = 1, 2, \ldots, K). \tag{2.46}$$

In these mass balances the chemical transformations have explicitly been taken into account with the knowledge of the stoichiometry and rates of the chemical reactions. Thus it is evident that the balances (2.45) and (2.46) are of fundamental importance in the theory of multicomponent and reacting continuous systems. It can also be said that these balance equations are perhaps the most important basic equations of chemical engineering sciences. It is remarkable that the integration of (2.43) leads to

$$c_k(t) = c_k(0) + \sum_{j=1}^{R} \nu_{kj}\xi_j(t), \qquad (k = 1, 2, \ldots, K) \tag{2.47}$$

where $c_k(0)$ are the initial values of the compositions at time $t = 0$. Since equation (1.32) is of general validity, at most $K - 1$ of these equations and ξ_j are independent.

3. Balances of Charge

It happens very often that some of the components in a multi-component and macroscopical continuous system have electric charge owing to the presence of corpuscular charge carriers (ions, electrons etc.). From the viewpoint of field theory no attention should be paid either to the corpuscular property of the charge carriers or to the discrete character of the microcharges, instead it is sufficient to indicate the charge of the individual components per unit mass. Since the electric charge is always bound to a corpuscle having mass, in the case of any k-th component its specific charge e_k, $(k = 1, 2, \ldots, K)$ referred to the unit mass, can be defined. The same is valid for the superposed continuum, the specific charge of which is

$$e = \varrho^{-1} \sum_{k=1}^{K} \varrho_k e_k = \sum_{k=1}^{K} c_k e_k. \tag{2.48}$$

Of course, for all the components which are uncharged, the specific charge is identically zero, i.e. $e_k \equiv 0$.

Let us determine the balance equations expressing the conservation of charge in the simple case when chemical reactions and charge exchange between the components are disregarded. The extension of the validity of the following equations to this more general case is very simple. In the absence of chemical reactions from (2.46) it follows that

$$\varrho \dot{c}_k + \nabla \cdot \boldsymbol{J}_k = 0, \qquad (k = 1, 2, \ldots, K) \tag{2.49}$$

from which the balance equations expressing the conservation of charge are immediately obtainable. First of all let us define the total electric current density $\boldsymbol{I}$ determined by the charge quantity carried by all components. This local current density is determined apart from the quantities e_k and ϱ_k by the individual velocities $\boldsymbol{v}_k$ of the components as follows

$$\boldsymbol{I} = \sum_{k=1}^{K} \varrho_k e_k \boldsymbol{v}_k = \varrho e \boldsymbol{v} + \sum_{k=1}^{K} e_k \boldsymbol{J}_k, \tag{2.50}$$

where (1.42) and (2.48) have been used. The quantity $\varrho e \boldsymbol{v}$ is the *convective* electric current density due to the convection of the barycenter with a velocity $\boldsymbol{v}$. The so-called *conductive* current density is determined by the last term of (2.50), i.e.

$$\boldsymbol{i} = \sum_{k=1}^{K} e_k \boldsymbol{J}_k \tag{2.51}$$

which is due to the motion of charged components relative to the center of mass. If the balance equation (2.49) is multiplied by the specific charge e_k of the k-th component and summing over all the components, then by using (2.50) and (2.51), we get the balance equation

$$\varrho \dot{e} + \nabla \cdot \boldsymbol{i} = 0 \tag{2.52}$$

which is the law of conservation of charge in substantial form.

Similarly the local form of the law of conservation of charge can be obtained from the zero reduced form of the local component balances (2.46):

$$\frac{\partial \varrho e}{\partial t} + \nabla \cdot \boldsymbol{I} = 0. \tag{2.53}$$

It is shown by the substantial (2.52) and local (2.53) charge balances, that the conductive current density $\boldsymbol{i}$ represents a substantial flux, whereas the total electric current density $\boldsymbol{I}$ represents a local flux. This can also be seen from the relation

$$\boldsymbol{i} = \boldsymbol{I} - \varrho e \boldsymbol{v} \tag{2.54}$$

which follows from definitions (2.50) and (2.51), furthermore from comparison with the general relation (2.12). At the same time, such a comparison also indicates that charge balances (2.52) and (2.53) could have been directly obtained from the general balances (2.15) and (2.8), respectively, by substitutions $a \equiv e$, $\boldsymbol{J}_a \equiv \boldsymbol{i}$, $\boldsymbol{J}_a^0 \equiv \boldsymbol{I}$ and $\sigma_a \equiv 0$. We note that the charge balances given here are of particular importance in thermo-electrodynamics as well as in electrochemistry and plasma physics.

4. The Equation of Motion

Before dealing with particular forms of impulse balances valid in the case of different system-models, let us derive their general form which is the equation of motion of continuum physics. Its original derivation relies upon the *stress principle* of CAUCHY [11]. The essence of this stress principle is that on an arbitrary surface element $\Delta\Omega_{\boldsymbol{n}}$ of external unit normal $\boldsymbol{n}$ belonging to every point $\boldsymbol{r}$ of a deformed continuum, stresses exist, which depend at any arbitrary moment on the place and the orientation of the surface element $\Delta\Omega_{\boldsymbol{n}}$, i.e. on the unit normal $\boldsymbol{n}$. The distribution of the stresses is described by the *stress vector*: $\boldsymbol{t}(\boldsymbol{r}, t; \boldsymbol{n})$. This vector can be interpreted as

$$\boldsymbol{t}_n(\boldsymbol{r}, t) = \lim_{\Delta\Omega_n \to 0} \frac{\Delta \boldsymbol{F}'}{\Delta\Omega_n} = \frac{\mathrm{d}\boldsymbol{F}'}{\mathrm{d}\Omega_n}, \tag{2.55}$$

where $\Delta\boldsymbol{F}'$ is the surface force which acts on $\Delta\Omega_n$. Hence the stress vector $\boldsymbol{t}_n$ is a typical field quantity, which also depends apart from place and time on the unit normal of the selected surface element (Fig. 3).

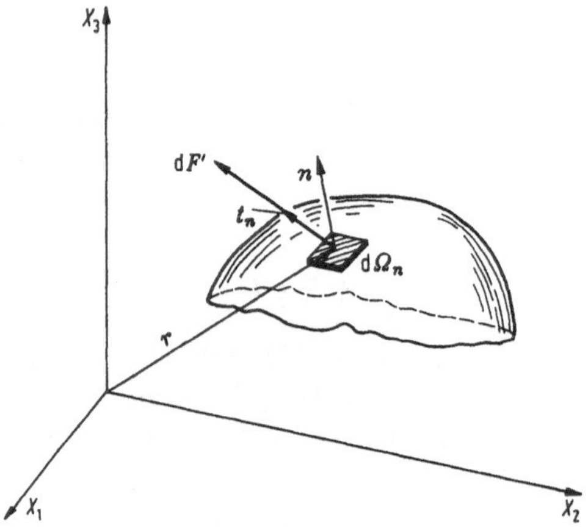

Fig. 3

Though from the viewpoint of field theory the existence of the limit value (2.55) is sufficient to derive the equation of motion of continuum physics, its essence, however, can be enlightened with the aid of the molecular viewpoints as well.

In an undeformed continuum, the arrangement of the corpuscles corresponds to the thermal equilibrium state at any time. In such a case the resultant of the forces acting upon any volume element of the continuum is zero. When the continuum is deformed, this equilibrium state is destroyed and internal stresses arise on each surface element $\mathrm{d}\Omega_n$ of the continuum which are locally determined per unit surface by the stress vector $\boldsymbol{t}_n$. These internal stresses arise from the interactions of the corpuscles (molecules, ions, atoms) and have a very short range, practically ranging only to the neighbouring particles. Hence, if a volume element $\mathrm{d}V$ is selected within the continuum it is only the surface force $\mathrm{d}\boldsymbol{F}'$ arising from interactions with the corpuscles present in the neighbouring volume elements, that acts upon it. The rectangular components of this force are determined by the components t_1, t_2, t_3

of the stress vector $\boldsymbol{t_n}$, i.e.

$$\mathrm{d}F'_\alpha = t_\alpha \,\mathrm{d}\Omega_{\boldsymbol{n}}, \qquad (\alpha = 1, 2, 3). \tag{2.56}$$

Hence $\mathrm{d}\boldsymbol{F}' = \boldsymbol{t_n}\,\mathrm{d}\Omega_{\boldsymbol{n}}$, is the surface force, which acts from the part of the neighbouring mass element $\mathrm{d}M' = \varrho\,\mathrm{d}V'$ upon the mass element $\mathrm{d}M = \varrho\,\mathrm{d}V$ along $\mathrm{d}\Omega_{\boldsymbol{n}}$ of an outward normal $\boldsymbol{n}$ and belongs to the volume element $\mathrm{d}V$. Since a surface element $\mathrm{d}\Omega_{\boldsymbol{n}}$ of outward normal $(-\boldsymbol{n})$ belongs to the neighbouring volume element $\mathrm{d}V'$ (Fig. 4),

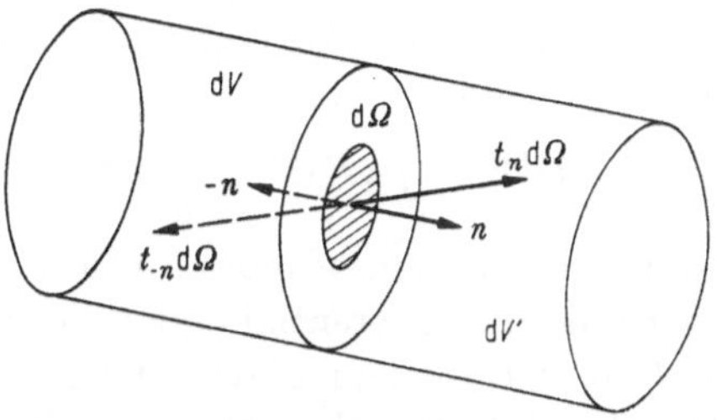

Fig. 4

$\mathrm{d}\boldsymbol{F}' = \boldsymbol{t}_{-\boldsymbol{n}}\,\mathrm{d}\Omega_{\boldsymbol{n}}$ is the force, with which the mass element $\mathrm{d}M = \varrho\,\mathrm{d}V$ acts upon the mass element $\mathrm{d}M' = \varrho\,\mathrm{d}V'$ along the surface $\mathrm{d}\Omega_{\boldsymbol{n}}$. According to the action-reaction law the two forces have equal magnitude but opposite direction, thus for the stress vector

$$\boldsymbol{t_n} = -\boldsymbol{t}_{-\boldsymbol{n}}, \tag{2.57}$$

is valid.

In order to determine the general equation of motion of continuum physics the Newtonian equation of motion of point mechanics must be generalized for continua. Consequently, the resultant $\boldsymbol{F}^*$ of all the forces acting upon all the material of the continuum of volume V must be given. This resultant force is determined in the general case by the surface forces (sometimes called stress forces or contact loads), further by the volume or external forces (sometimes called extraneous or extrinsic forces). The latter act upon each part of the continuum (upon the internal part) hence they are proportional to the mass incorporated in the system and to the volume of the system, respectively. Such a typical force is, for instance, the gravitational force which is in case of a mass element $\mathrm{d}M = \varrho\,\mathrm{d}V$ proportional to the mass $\mathrm{d}M$ and the volume $\mathrm{d}V$, respectively. In general, the external force fields act with such a force on the mass elements of the continuum. If the external force acting upon the mass element $\mathrm{d}M$ is $\mathrm{d}\boldsymbol{F}''$, then

$$\mathrm{d}\boldsymbol{F}'' = \boldsymbol{F}\,\mathrm{d}M = \varrho\boldsymbol{F}\,\mathrm{d}V, \tag{2.58}$$

is valid where $\boldsymbol{F}$ is the external force per unit mass. With (2.56) and (2.58) the resultant of the elementary surface and volume forces can be given by

$$\mathrm{d}\boldsymbol{F}^* = \varrho\boldsymbol{F}\,\mathrm{d}V + \oint_{\mathrm{d}\Omega} \boldsymbol{t_n}\,\mathrm{d}\Omega_{\boldsymbol{n}} \tag{2.59}$$

where the surface integral must be extended to the $d\Omega$ boundary surface of the volume dV. Consequently, if $\dot{\boldsymbol{v}}$ is the acceleration of the mass element dM which can be ascribed to the action of the force $d\boldsymbol{F}$, the Newtonian equation of motion of point mechanics for the mass element dM as a "particle" must be written in the form

$$\varrho \dot{\boldsymbol{v}}\, dV = \varrho \boldsymbol{F}\, dV + \oint_{d\Omega} \boldsymbol{t}_{\boldsymbol{n}}\, d\Omega_{\boldsymbol{n}}. \tag{2.60}$$

The global form of the equation of motion of the continuum is obtained by integration of this expression extended to the total volume V and boundary surface Ω of the continuum. Hence

$$\int_V \varrho \dot{\boldsymbol{v}}\, dV = \int_V \varrho \boldsymbol{F}\, dV + \oint_\Omega \boldsymbol{t}\, d\Omega \tag{2.61}$$

is the global form of the equation of motion of the continuum.

Now let us determine the differential equation valid for each point of the continuum. Consider an elementary tetrahedron limited by the surfaces $d\Omega_{\boldsymbol{n}}$, $d\Omega_{\boldsymbol{i}}$, $d\Omega_{\boldsymbol{j}}$, $d\Omega_{\boldsymbol{k}}$ and the outward unit normals of which point towards the directions: $\boldsymbol{n}$, $-\boldsymbol{i}$, $-\boldsymbol{j}$, $-\boldsymbol{k}$ (cf. Fig. 5). The surface

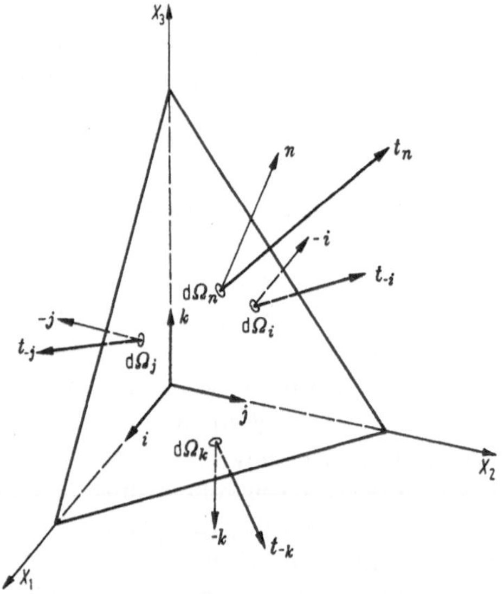

Fig. 5

forces acting upon the faces of the tetrahedron are $d\boldsymbol{F}' = \boldsymbol{t}_{\boldsymbol{n}}\, d\Omega_{\boldsymbol{n}}$ and $d\boldsymbol{F}'_{\boldsymbol{\alpha}} = \boldsymbol{t}_{-\boldsymbol{\alpha}}\, d\Omega_{\boldsymbol{\alpha}}$, $(\boldsymbol{\alpha} = \boldsymbol{i}, \boldsymbol{j}, \boldsymbol{k})$. Any elementary tetrahedron taken at an arbitrary place of the continuum will be in equilibrium due to the common action of these forces, if the resultant of the four forces vanishes, i.e. the local condition

$$\boldsymbol{t}_{\boldsymbol{n}}\, d\Omega_{\boldsymbol{n}} + \boldsymbol{t}_{-\boldsymbol{i}}\, d\Omega_{\boldsymbol{i}} + \boldsymbol{t}_{-\boldsymbol{j}}\, d\Omega_{\boldsymbol{j}} + \boldsymbol{t}_{-\boldsymbol{k}}\, d\Omega_{\boldsymbol{k}} = 0 \tag{2.62}$$

is fulfilled. According to the action-reaction law (2.57) however, the relations

$$\boldsymbol{t}_i = -\boldsymbol{t}_{-i}, \quad \boldsymbol{t}_j = -\boldsymbol{t}_{-j}, \quad \boldsymbol{t}_k = -\boldsymbol{t}_{-k}, \tag{2.63}$$

are valid and thus the condition (2.62) leads with the use of (2.63) and the relations $\mathrm{d}\Omega_\alpha = n_\alpha \, \mathrm{d}\Omega_n$, $(\alpha = \boldsymbol{i}, \boldsymbol{j}, \boldsymbol{k})$ to

$$\boldsymbol{t}_n = n_i \boldsymbol{t}_i + n_j \boldsymbol{t}_j + n_k \boldsymbol{t}_k, \tag{2.64}$$

where n_i, n_j, n_k are the rectangular components of the unit normal $\boldsymbol{n}$ which are determined by the direction cosinus of the vector $\boldsymbol{n}$. The equation (2.64) means that of the stress vector $\boldsymbol{t}_n$ is a homogeneous linear vector function of the unit normal $\boldsymbol{n}$, i.e. for any point of the continuum equation

$$t_\beta(\boldsymbol{r}, t) = n_\alpha T_{\alpha\beta}(\boldsymbol{r}, t), \qquad (\alpha, \beta = 1, 2, 3), \tag{2.65}$$

is valid where the matrix of nine scalar quantities $T_{\alpha\beta}$ forms a second order tensor

$$\mathbf{T}(\boldsymbol{r}, t) \equiv [T_{\alpha\beta}] = \begin{bmatrix} T_{11} & T_{21} & T_{31} \\ T_{12} & T_{22} & T_{32} \\ T_{13} & T_{23} & T_{33} \end{bmatrix} \tag{2.66}$$

which is called the *stress tensor*. Equation (2.65) which may be written also in a tensor form

$$\boldsymbol{t} = \boldsymbol{n} \cdot \mathbf{T} \tag{2.67}$$

states that *if the stresses determined by the elements $T_{\alpha\beta}$ of the stress tensor* $\mathbf{T}$ *in a given place of the continuum are known, any stress vector can be determined in the same place.* The three elements $T_{\alpha\alpha}$ of the tensor $\mathbf{T}$, if they are positive, are called normal stresses or tensions, whereas if they are negative, they are called normal pressures. The elements $T_{\alpha\beta}$, $\alpha \neq \beta$, are tangential or shearing stresses respectively.

The differential equation representing the equation of motion of the continuum can immediately be derived from (2.61). Substituting the stress vector from (2.67) into (2.61) and applying the Gauss divergence theorem, we have

$$\int_V \varrho \dot{\boldsymbol{v}} \, \mathrm{d}V = \int_V (\varrho \boldsymbol{F} + \nabla \cdot \mathbf{T}) \, \mathrm{d}V, \tag{2.68}$$

from which, since V is arbitrary, the equation of motion follows

$$\varrho \dot{\boldsymbol{v}} = \varrho \boldsymbol{F} + \nabla \cdot \mathbf{T} \tag{2.69}$$

originally due to Cauchy. In this compact and elegant equation by the symbol $\nabla \cdot \mathbf{T}$ a tensor divergence is denoted and $\dot{\boldsymbol{v}}$ is the barycentric acceleration because the substantial time derivative of the barycentric velocity occurs in it. The equation of motion (2.69) is valid in the case of any continuum, but of course the expression of the stress tensor is different for the different continuum-models. The different models of deformable bodies (elastic, plastic, etc.), the hydrodynamic models (ideal and viscous fluids, turbulent systems, etc.) moreover the various system-models of electromagnetic fields can also be treated by the ekuation of motion (2.69), if the actual forms of the stress tensor are qnown.

5. Balances of Impulse

Equations (2.61) and (2.68) express the impulse conservation in a global form: *the variation of the impulse of a material volume in the course of time is equal to the resultant of the forces acting upon the substance present in the volume.* Hence, the physical content of differential equation (2.69) is also identical with the conservation of impulse. This is particularly evident from another form of the equation of motion, which is even more frequently used in thermodynamics than (2.69). The application of (2.69) is more familiar in mechanics. The alternative form of the equation of motion is obtained, if the pressure tensor interpreted as the negative of the stress tensor

$$\mathbf{P} \equiv -\mathbf{T} \tag{2.70}$$

is introduced, by which (2.69) can be rewritten as

$$\varrho \dot{\boldsymbol{v}} + \nabla \cdot \mathbf{P} = \varrho \boldsymbol{F}. \tag{2.71}$$

Comparing this equation to the general substantial balance (2.15) and remembering that $\boldsymbol{v}$ is the specific impulse, it can be seen that (2.71) is an impulse balance in substantial form. Indeed, (2.71) is a balance equation to which the substantial impulse current density identical with the second order pressure tensor

$$\mathbf{J}_{\text{imp}} = \mathbf{P} \tag{2.72}$$

and the source term

$$\sigma^{\text{k}}_{\text{imp}} = \varrho \boldsymbol{F} \tag{2.73}$$

determined by the external force density belongs. The force density $\varrho \boldsymbol{F}$ can evidently be interpreted only as an "external" source term. Namely, this source of the impulse has its real cause in the external fields acting on the continuum. Thus, in spite of the fact that the equation of motion (2.71) is a balance equation which has a source-term, it expresses the conservation of impulse in a differential form. The source-term arising from local inhomogeneities inside the system does not occur in the balance (2.71), i.e. $\sigma^{\text{i}}_{\text{imp}} \equiv 0$.

The substantial balance (2.71) can be transformed with (2.21) into the local form

$$\frac{\partial \varrho \boldsymbol{v}}{\partial t} + \nabla \cdot (\mathbf{P} + \varrho \boldsymbol{v}\boldsymbol{v}) = \varrho \boldsymbol{F}, \tag{2.74}$$

if in (2.21) under the quantity a the components of the barycentric velocity, i.e. $a \equiv v_\alpha$ $(\alpha = 1, 2, 3)$ are understood. Consequently, in (2.74) $\boldsymbol{vv}$ means a dyadic product, whereas the local impulse current density

$$\mathbf{J}^0_{\text{imp}} = \mathbf{P} + \varrho \boldsymbol{v}\boldsymbol{v} \tag{2.75}$$

is composed of the $\mathbf{P}$ substantial (conductive), and the convective part $\varrho \boldsymbol{vv}$, in agreement with the general expression (2.12).

Let us now examine the problem: which are the conditions to ensure that the impulse balances (2.71) and (2.74) can be considered as the impulse balances of a continuum which was obtained as the local superposition of K continuous constituents. First of all let us give the impulse balance analogous to (2.71) for the k-th component medium. This can be given as

$$\varrho_k \frac{\mathrm{d}^{(k)} \boldsymbol{v}_k}{\mathrm{d}t} + \nabla \cdot \mathbf{P}_k = \varrho_k \boldsymbol{F}_k + \varrho_k \boldsymbol{F}^*_k, \qquad (k = 1, 2, \ldots, K) \tag{2.76}$$

where the time derivative of the component velocity $\boldsymbol{v}_k$ is, according to (1.44), the substantial time derivative referred to the k-th medium. In (2.76) $\mathbf{P}_k$ is the pressure tensor of the k-th continuum, $\boldsymbol{F}_k$ the external force acting upon the unit mass of the k-th component, whereas $\boldsymbol{F}^*_k$ can be interpreted as an internal force, which arises from the other components and—according to the corpuscular theory—acts upon the unit mass of the k-th component as a resultant of the intermolecular short-range forces [12, 13]. Anyway, the generality is only extended by assuming that the impulse balances (2.76) referring to the individual components include internal forces $\boldsymbol{F}^*_k$ which are accompanied by the "internal" impulse source $\boldsymbol{\sigma}^{\mathrm{i}}_{\text{imp}} = \varrho_k \boldsymbol{F}^*_k$.

In order to determine the conditions by which it is ensured that balance equation (2.71) be valid also in case of multi-component systems, let us apply for the elimination of $\frac{\mathrm{d}^{(k)}}{\mathrm{d}t}$ from (2.76) the formula (1.45) with $A \equiv v_{k\alpha}$ $(\alpha = 1, 2, 3)$. We obtain

$$\frac{\mathrm{d}^{(k)} \boldsymbol{v}_k}{\mathrm{d}t} = \dot{\boldsymbol{v}}_k + (\boldsymbol{w}_k \cdot \nabla) \boldsymbol{v}_k, \qquad (k = 1, 2, \ldots, K) \tag{2.77}$$

by which (2.76) may be rewritten as

$$\varrho_k \dot{\boldsymbol{v}}_k + (\boldsymbol{J}_k \cdot \nabla) \boldsymbol{v}_k + \nabla \cdot \boldsymbol{P}_k = \varrho_k \boldsymbol{F}_k + \varrho_k \boldsymbol{F}^*_k, \; (k = 1, 2, \ldots, K). \tag{2.78}$$

This equation can be transformed with the identity

$$\nabla \cdot (\boldsymbol{J}_k \boldsymbol{v}_k) = \boldsymbol{v}_k (\nabla \cdot \boldsymbol{J}_k) + (\boldsymbol{J}_k \cdot \nabla) \boldsymbol{v}_k$$

into

$$\varrho_k \dot{\boldsymbol{v}}_k + \nabla \cdot (\boldsymbol{P}_k + \boldsymbol{v}_k \boldsymbol{J}_k) = \varrho_k \boldsymbol{F}_k + \varrho_k \boldsymbol{F}_k^* + \boldsymbol{v}_k (\nabla \cdot \boldsymbol{J}_k), \quad (k = 1, 2, \ldots, K) \tag{2.79}$$

which can be written after summation and with the help of the substantial time derivative of (1.39a), i.e. of the relation

$$\sum_{k=1}^{K} \varrho_k \dot{\boldsymbol{v}}_k = \varrho \dot{\boldsymbol{v}} + \dot{\varrho} \boldsymbol{v} - \sum_{k=1}^{K} \dot{\varrho}_k \boldsymbol{v}_k \tag{2.80}$$

in the following form:

$$\begin{aligned} &\dot{\varrho} \boldsymbol{v} + \varrho \dot{\boldsymbol{v}} - \sum_{k=1}^{K} \dot{\varrho}_k \boldsymbol{v}_k + \nabla \cdot \left\{ \sum_{k=1}^{K} (\mathbf{P}_k + \boldsymbol{J}_k \boldsymbol{v}_k) \right\} \\ &= \sum_{k=1}^{K} \varrho_k \boldsymbol{F}_k + \sum_{k=1}^{K} \varrho_k \boldsymbol{F}_k^* + \sum_{k=1}^{K} \boldsymbol{v}_k (\nabla \cdot \boldsymbol{J}_k). \end{aligned} \tag{2.81}$$

Applying now for the elimination of $\dot{\varrho}$ and $\dot{\varrho}_k$ the balances (2.17) and (2.33), further, taking into account that owing to (1.41), (1.42) and (1.43)

$$\begin{aligned} \sum_{k=1}^{K} \boldsymbol{J}_k \boldsymbol{v}_k &= \sum_{k=1}^{K} \varrho_k (\boldsymbol{v}_k - \boldsymbol{v}) \boldsymbol{v}_k = \sum_{k=1}^{K} \varrho_k \boldsymbol{w}_k \boldsymbol{v}_k \\ &= \sum_{k=1}^{K} \varrho_k \boldsymbol{w}_k (\boldsymbol{v}_k - \boldsymbol{v}) + \sum_{k=1}^{K} \varrho_k \boldsymbol{w}_k \boldsymbol{v} = \sum_{k=1}^{K} \varrho_k \boldsymbol{w}_k \boldsymbol{w}_k \end{aligned} \tag{2.82}$$

is valid, and after a simple calculation (2.81) yields the balance equation

$$\varrho \dot{\boldsymbol{v}} + \nabla \cdot \left\{ \sum_{k=1}^{K} (\mathbf{P}_k + \varrho_k \boldsymbol{w}_k \boldsymbol{w}_k) \right\} = \sum_{k=1}^{K} \varrho_k \boldsymbol{F}_k + \sum_{k=1}^{K} (\varrho_k \boldsymbol{F}_k^* + m_k \boldsymbol{v}_k). \tag{2.83}$$

Comparing this equation derived originally by Truesdell in a somewhat different way [1, 14] with the Cauchy equation (2.71), it can be seen that the necessary and sufficient condition for the equation of Cauchy to be valid for multi-component continua also is equivalent to the satisfaction of the following three requirements:

$$\boldsymbol{F} = \varrho^{-1} \sum_{k=1}^{K} \varrho_k \boldsymbol{F}_k = \sum_{k=1}^{K} c_k \boldsymbol{F}_k \tag{2.84}$$

and

$$\mathbf{P} = \sum_{k=1}^{K} (\mathbf{P}_k + \varrho_k \boldsymbol{w}_k \boldsymbol{w}_k), \tag{2.85}$$

and finally

$$\sum_{k=1}^{K} (\varrho_k \boldsymbol{F}_k^* + m_k \boldsymbol{v}_k) = 0. \tag{2.86}$$

This latter condition asserts that the impulse source of the internal forces is balanced by the diffusion impulse of the components formed in the chemical reactions. Consequently, the impulse balance valid in multi-component continua

$$\varrho \dot{\boldsymbol{v}} + \nabla \cdot \mathbf{P} = \sum_{k=1}^{K} \varrho_k \boldsymbol{F}_k \tag{2.87}$$

expresses the conservation of the total impulse in the case of a pressure tensor determined by (2.85).

Cauchy's equation of motion of continuum physics can be applied to most diversive cases of continuum-models of mechanics as well as to electromagnetic continua. The various continuum-models differ by the forms of the pressure tensor $\mathbf{P}$, which can be verified – though in an indirect way – experimentally. Examining continua of mechanical character only, (more precise being interested if the pressure tensor represents the mechanical properties of a continuum) $\mathbf{P}$ can in general be splitted into two parts. The first part depends on the state, whereas the other on the velocity of the state variation. This means that the mechanical pressure tensor $\mathbf{P}$ includes an equilibrium part $\mathbf{P}^e$ and a non-equilibrium part $\mathbf{P}^v$, i.e.

$$\mathbf{P} = \mathbf{P}^e + \mathbf{P}^v . \tag{2.88}$$

Since $\mathbf{P}^v$ depends on the velocity of the state variation and on its gradient respectively, and because the viscous forces are determined by the latter, $\mathbf{P}^v$ is called the *viscous pressure tensor*. Hence, in the case of the equilibrium of any continuum, $\mathbf{P}^v = 0$. For example, in the case of gases or fluids at rest the non-equilibrium pressure tensor is zero, furthermore, in such cases, owing to the isotropy, the equilibrium pressure tensor $\mathbf{P}^e$ is reduced to a scalar, i.e.

$$\mathbf{P} = \mathbf{P}^e = p\,\boldsymbol{\delta}, \tag{2.89}$$

where p is the scalar hydrostatic pressure and $\boldsymbol{\delta}$ is the unit tensor. The reduction of $\mathbf{P}$ given in detail for the scalar pressure

$$\mathbf{P} = \mathbf{P}^e = \begin{bmatrix} p & 0 & 0 \\ 0 & p & 0 \\ 0 & 0 & p \end{bmatrix} \tag{2.90}$$

shows that in the case of fluids and gases at rest the shearing pressure (or shearing tensions) determined by the elements of mixed indices of the pressure tensor do not occur, i.e. $P_{\alpha\beta} = 0$, $(\alpha, \beta = 1, 2, 3; \alpha \neq \beta)$. On the other hand, in such continua the normal pressures (negative normal tensions) are equal to one another and the hydrostatic pressure

$$P_{\alpha\alpha} = P^e_{\alpha\alpha} = p, \qquad (\alpha = 1, 2, 3) \tag{2.91}$$

is determined by them. Expression (2.91) expresses the isotropy of hydrostatic pressure, which is based upon the Pascal law derived from experiments.

Relations (2.89) to (2.91) are always valid for fluids and gases at rest. However, in the case of moving fluids and gas systems system-models can also be formulated, for which the relations in question represent a good approximation. Fluid systems where the tangential tensions do not occur are called viscousless (perfect or ideal) systems, by definition. Hence, the expression of the pressure tensor (2.89) also determines the model of the ideal fluid and gas systems. If the relation (2.89) in the equation of motion (2.71) is taken into account, the substantial impulse balance valid in the case of ideal systems

$$\varrho \dot{\boldsymbol{v}} + \nabla p = \varrho \boldsymbol{F} \tag{2.92}$$

or, after the application of (1.15) to the velocity components, the local impulse balance

$$\frac{\partial \boldsymbol{v}}{\partial t} + (\boldsymbol{v} \cdot \nabla)\, \boldsymbol{v} + \varrho^{-1}\, \nabla p = \boldsymbol{F} \tag{2.93}$$

is obtained. These equations are known as the *Euler equations of motion of hydrodynamics*, and in the sense of what has been said before, these are the impulse balances of viscousless fluid systems.

In case of viscous systems the form (2.89) or (2.90) of the pressure tensor is modified, because in such cases the viscous pressure tensor is not zero. For homogeneous isotropic viscous systems the form of the total mechanical pressure tensor is

$$\mathbf{P} = p\, \boldsymbol{\delta} + \mathbf{P}^{\mathrm{V}}, \tag{2.94}$$

where $\mathbf{P}^{\mathrm{V}}$ is, in general, a function of the tensor $\nabla \boldsymbol{v} = \operatorname{Grad} \boldsymbol{v}$. If we restrict our considerations to the Newtonian fluids for which it is characteristic that the substantial impulse current density determined by the viscous pressure tensor $\mathbf{P}^{\mathrm{V}}$ is a linear function of the tensor $\nabla \boldsymbol{v}$, expression

$$\mathbf{P}^{\mathrm{V}} = \left\{\left(\frac{2}{3}\eta - \eta_{\mathrm{V}}\right) \nabla \cdot \boldsymbol{v}\right\} \boldsymbol{\delta} - 2\eta (\nabla \boldsymbol{v})^{\mathrm{s}} \tag{2.95}$$

is valid, which is the so-called Newtonian viscous pressure tensor. Here η is the shear viscosity and η_{V} is the coefficient of volume (bulk) viscosity, whereas $(\nabla \boldsymbol{v})^{\mathrm{s}}$ is the symmetric part of the velocity gradient tensor $\nabla \boldsymbol{v}$. With (2.94) and (2.95) the total Newtonian pressure tensor

$$\mathbf{P} = \left\{p + \left(\frac{2}{3}\eta - \eta_{\mathrm{V}}\right) \nabla \cdot \boldsymbol{v}\right\} \boldsymbol{\delta} - 2\eta\, (\nabla \boldsymbol{v})^{\mathrm{s}} \tag{2.96}$$

is obtained. Substituting this into (2.71) and after having carried out the indicated divergence, the Navier-Stokes equation is obtained in substantial form:

$$\varrho \dot{\boldsymbol{v}} + \nabla p - \eta \, \Delta \boldsymbol{v} - \left(\frac{\eta}{3} + \eta_{\mathrm{v}}\right) \nabla\nabla \cdot \boldsymbol{v} = \varrho \boldsymbol{F}. \tag{2.97}$$

From this, by using (1.15), we have its local form

$$\varrho \frac{\partial \boldsymbol{v}}{\partial t} + \varrho(\boldsymbol{v} \cdot \nabla) \boldsymbol{v} + \nabla p - \eta \, \Delta \boldsymbol{v} - \left(\frac{\eta}{3} + \eta_{\mathrm{v}}\right) \nabla\nabla \cdot \boldsymbol{v} = \varrho \boldsymbol{F}. \tag{2.98}$$

In these equations the Laplace operator is denoted by $\Delta \equiv \nabla^2$. The given equations of motion must be considered as the impulse balances in the substantial and in the local form, respectively, of the viscous Newtonian fluids.

For the sake of the subsequent analysis let us split up the viscous pressure tensor (2.88) in the following way

$$\mathbf{P}^{\mathrm{v}} = p^{\mathrm{v}} \boldsymbol{\delta} + \overset{0}{\mathbf{P}}{}^{\mathrm{v}} \tag{2.99}$$

where $\boldsymbol{\delta}$ is the unit tensor and

$$p^{\mathrm{v}} = \frac{1}{3} \mathbf{P}^{\mathrm{v}} : \boldsymbol{\delta} = \frac{1}{3} P^{\mathrm{v}}_{\alpha\alpha}, \qquad (\alpha = 1, 2, 3) \tag{2.100}$$

is one third of the trace of the tensor $\mathbf{P}^{\mathrm{v}}$. According to (2.99) the trace of the tensor $\overset{0}{\mathbf{P}}{}^{\mathrm{v}}$ is zero, i.e.

$$\overset{0}{\mathbf{P}}{}^{\mathrm{v}} : \boldsymbol{\delta} = \overset{0}{P}{}^{\mathrm{v}}_{\alpha\alpha} = 0, \qquad (\alpha = 1, 2, 3). \tag{2.101}$$

This tensor can be split into a symmetric $\overset{0}{\mathbf{P}}{}^{\mathrm{vs}}$ and an antisymmetric part $\mathbf{P}^{\mathrm{va}}$ in the usual way ($\mathbf{P}^{\mathrm{va}}$ has zero trace, by definition):

$$\overset{0}{\mathbf{P}}{}^{\mathrm{v}} = \overset{0}{\mathbf{P}}{}^{\mathrm{vs}} + \mathbf{P}^{\mathrm{va}}. \tag{2.102}$$

Hence, the total viscous pressure tensor is according to the maximal decomposition of the following form

$$\mathbf{P}^{\mathrm{v}} = p^{\mathrm{v}} \boldsymbol{\delta} + \overset{0}{\mathbf{P}}{}^{\mathrm{vs}} + \mathbf{P}^{\mathrm{va}} = \frac{1}{3} (\mathbf{P}^{\mathrm{v}} : \boldsymbol{\delta}) \, \boldsymbol{\delta} + \overset{0}{\mathbf{P}}{}^{\mathrm{vs}} + \mathbf{P}^{\mathrm{va}}. \tag{2.103}$$

This decomposition can be carried out in the case of every second order tensor and because it will be applied in other cases as well it can be written with components:

$$\begin{bmatrix} P^{\mathrm{v}}_{11} & P^{\mathrm{v}}_{21} & P^{\mathrm{v}}_{31} \\ P^{\mathrm{v}}_{12} & P^{\mathrm{v}}_{22} & P^{\mathrm{v}}_{32} \\ P^{\mathrm{v}}_{13} & P^{\mathrm{v}}_{23} & P^{\mathrm{v}}_{33} \end{bmatrix} = \frac{P^{\mathrm{v}}_{11} + P^{\mathrm{v}}_{22} + P^{\mathrm{v}}_{33}}{3} \begin{bmatrix} 1 & 0 & 0 \\ 0 & 1 & 0 \\ 0 & 0 & 1 \end{bmatrix}$$

$$+ \begin{bmatrix} \frac{2P^{\mathrm{v}}_{11} - P^{\mathrm{v}}_{22} - P^{\mathrm{v}}_{33}}{3} & \frac{P^{\mathrm{v}}_{21} + P^{\mathrm{v}}_{12}}{2} & \frac{P^{\mathrm{v}}_{31} + P^{\mathrm{v}}_{13}}{2} \\ \frac{P^{\mathrm{v}}_{12} + P^{\mathrm{v}}_{21}}{2} & \frac{2P^{\mathrm{v}}_{22} - P^{\mathrm{v}}_{11} - P^{\mathrm{v}}_{33}}{3} & \frac{P^{\mathrm{v}}_{32} + P^{\mathrm{v}}_{23}}{2} \\ \frac{P^{\mathrm{v}}_{13} + P^{\mathrm{v}}_{31}}{2} & \frac{P^{\mathrm{v}}_{23} + P^{\mathrm{v}}_{32}}{2} & \frac{2P^{\mathrm{v}}_{33} - P^{\mathrm{v}}_{11} - P^{\mathrm{v}}_{22}}{3} \end{bmatrix} \tag{2.104}$$

$$+ \begin{bmatrix} 0 & \frac{P^{\mathrm{v}}_{21} - P^{\mathrm{v}}_{12}}{2} & \frac{P^{\mathrm{v}}_{31} - P^{\mathrm{v}}_{13}}{2} \\ \frac{P^{\mathrm{v}}_{12} - P^{\mathrm{v}}_{21}}{2} & 0 & \frac{P^{\mathrm{v}}_{32} - P^{\mathrm{v}}_{23}}{2} \\ \frac{P^{\mathrm{v}}_{13} - P^{\mathrm{v}}_{31}}{2} & \frac{P^{\mathrm{v}}_{23} - P^{\mathrm{v}}_{32}}{2} & 0 \end{bmatrix}.$$

It can be seen from this decomposition that in the general case a physical quantity determined by nine independent elements of a second order tensor can be determined in an equivalent manner by a scalar equal to one third of the trace of the tensor, by five independent elements of the symmetric part of zero trace, and by the three independent elements of the antisymmetric part. The latter is equivalent to the axial vector formed by cyclical permutations of the respective tensor elements, viz.

$$P^{\mathrm{va}}_{\alpha\beta} = -P^{\mathrm{va}}_{\beta\alpha} = P^{\mathrm{va}}_{\gamma}, \qquad (\alpha, \beta, \gamma = 1, 2, 3) \tag{2.105}$$

where

$$P^{\mathrm{va}}_{\alpha\beta} = \frac{P^{\mathrm{v}}_{\alpha\beta} - P^{\mathrm{v}}_{\beta\alpha}}{2}, \qquad (\alpha, \beta = 1, 2, 3). \tag{2.106}$$

The expression (2.90) or (2.91) defines, in the strict sense, only the hydrostatic pressure. We arrive at a more general definition of the pressure (valid also in the non-static case), if this is interpreted as one third of the trace of the total pressure tensor:

$$\bar{p} = \frac{1}{3}\mathbf{P} : \boldsymbol{\delta} = \frac{1}{3} P_{\alpha\alpha}, \qquad (\alpha = 1, 2, 3). \tag{2.107}$$

This quantity is the mean hydrodynamic (or rather thermodynamic) pressure $\bar{p}$ which is composed according to (2.88) of the equilibrium pressure

$$p = \frac{1}{3}\mathbf{P}^{\mathrm{e}} : \boldsymbol{\delta} = \frac{1}{3} P^{\mathrm{e}}_{\alpha\alpha}, \qquad (\alpha = 1, 2, 3) \tag{2.108}$$

and of the scalar p^{v} defined in (2.100). Consequently, p^{v} may be called *viscous pressure*. It follows from (2.88) with (2.103) and (2.108) that for moving fluid and gas systems the pressure tensor is

$$\mathbf{P} = (p + p^{\text{v}})\,\boldsymbol{\delta} + \overset{0}{\mathbf{P}}{}^{\text{vs}} + \mathbf{P}^{\text{va}}, \tag{2.109}$$

which will be used later on in this form.

Pressure tensor expressions representing other continua-models will not be given. We note, however, that by a slight generalization of the Newtonian pressure tensor (2.96), the Reynolds pressure tensor of turbulent motion and the Reynolds equations of motion is also obtainable [15]. It is also worthwhile to note that with the various forms of the equilibrium part $\mathbf{P}^{\text{e}}$ of the pressure tensor the models of elastic, plastic and rheological systems can be described unambiguously and the agreement with experimental facts is good. Though all the system-models are of fundamental importance from the viewpoints of physicists, rheologists and chemists of plastics dealing with thermomechanics their treatment does not fall within the scope of this treatise. The primary reason for this is that the systematic applications of non-equilibrium thermodynamics to thermomechanical and rheological systems was started only some years before. We refer to Kluitenberg's fundamental works [16].

6. The Mechanical Equilibrium

The so-called mechanical equilibrium states of continua are equally important from practical and thoretical points of view. According to mechanics it is characteristic for the mechanical equilibrium state of any body that in such a state the body cannot carry out either a translational motion (with the exception of the straight uniform translation) or a rotational one. This condition, i.e. a body is in mechanical equilibrium, is satisfied if the resultant of the forces acting upon any volume of the body as well as the resultant of the moments of all the forces are zero.

The first condition in case of continua is fulfilled according to (2.68), (putting instead of $\mathbf{T}$ the pressure tensor $-\mathbf{P}$), if

$$\int_V (\varrho \boldsymbol{F} - \nabla \cdot \mathbf{P})\,\mathrm{d}V = 0. \tag{2.110}$$

Since V is arbitrary, and in equilibrium at each point of the continuum equation

$$\varrho \boldsymbol{F} = \nabla \mathbf{P} \tag{2.111}$$

expressing the equilibrium condition must hold. According to (2.71) by this condition an accelerationless motion is required, i.e. in the state of mechanical equilibrium $\dot{\boldsymbol{v}} = 0$.

By the condition (2.111) it is only the accelerating translational motion whose possibility is excluded, but the rotational motion of the mass elements of continua is allowed. Consequently this is the necessary condition for a total mechanical equilibrium state. The necessary and sufficient condition of the mechanical equilibrium is obtained, if apart from equation (2.111) the condition expressing the vanishing of the momenta of forces is also given. Since in the case of continua the momentum of the resultant force is defined with respect to the point $\boldsymbol{r}$ of the spatial system of coordinates as

$$\int_V \varrho(\boldsymbol{r}\times\boldsymbol{F})\,\mathrm{d}V + \oint_\Omega (\boldsymbol{r}\times\boldsymbol{t})\,\mathrm{d}\Omega = \int_V \varrho(\boldsymbol{r}\times\boldsymbol{F})\,\mathrm{d}V - \oint_\Omega (\boldsymbol{r}\times\mathbf{P})\cdot\mathrm{d}\boldsymbol{\Omega} \quad (2.112)$$

(where relations (2.67) and (2.70) have been used on the right-hand side), and thus the second condition of the mechanical equilibrium

$$\int_V [\varrho(\boldsymbol{r}\times\boldsymbol{F}) - \nabla\cdot(\boldsymbol{r}\times\mathbf{P})]\,\mathrm{d}V = 0 \quad (2.113)$$

is obtained by the application of the divergence theorem to the last term of (2.112). Making us of the tensor analytical identity

$$x_\alpha \frac{\partial P_{\gamma\beta}}{\partial x_\gamma} - x_\beta \frac{\partial P_{\gamma\alpha}}{\partial x_\gamma} = \frac{\partial}{\partial x_\gamma}(x_\alpha P_{\gamma\beta} - x_\beta P_{\gamma\alpha}) + P_{\beta\alpha} - P_{\alpha\beta}, \quad (\alpha, \beta, \gamma = 1, 2, 3) \quad (2.114)$$

in the vector form

$$\boldsymbol{r}\times\nabla\cdot\mathbf{P} = \nabla\cdot(\boldsymbol{r}\times\mathbf{P}) + \tilde{\mathbf{P}} - \mathbf{P} \quad (2.115)$$

(where $\tilde{\mathbf{P}}$ is the transposed tensor of $\mathbf{P}$), the condition (2.113) can be rewritten as

$$\int_V [\boldsymbol{r}\times(\varrho\boldsymbol{F} - \nabla\cdot\mathbf{P}) + \tilde{\mathbf{P}} - \mathbf{P}]\,\mathrm{d}V = 0. \quad (2.116)$$

Since the integrand must vanish for an arbitrary volume the simultaneous fulfilment of (2.111) and (2.113) expresses the necessary and sufficient condition for mechanical equilibrium, and leads to

$$\mathbf{P} = \tilde{\mathbf{P}} \qquad \text{i.e.}\ P_{\alpha\beta} = P_{\beta\alpha}, \qquad (\alpha, \beta = 1, 2, 3) \quad (2.117)$$

i.e. that in case of mechanical equilibrium the pressure tensor is symmetric. In other cases the symmetry of the pressure tensor is related to the conservation of the angular momentum. Before analysing in detail the balance equations valid for the angular momentum, a concept of mechanical equilibrium often used should be mentioned which can be formulated under less strict assumptions than the ones given hitherto.

These conditions are often practically fulfilled in multi-component hydro-thermodynamic systems with good approximation it has been primarily used by Prigogine [17]. The Prigogine mechanical equilibrium refers only to a multi-component and viscousless continua. On the other hand, even in such cases it is restricted only to condition (2.111).

Let us consider a multi-component hydrodynamic continuum whose equation of motion is represented by (2.87). Prigogine's mechanical equilibrium is characterized now not only by $\dot{\boldsymbol{v}} = 0$, but also by $\mathbf{P}^{\mathrm{v}} = f(\nabla \boldsymbol{v}) = 0$. In this sense the Prigogine mechanical equilibrium is expressed by the condition

$$\nabla p = \sum_{k=1}^{K} \varrho_k \boldsymbol{F}_k \qquad (2.118)$$

which has a very important role in the description of diffusion, etc. For example, with diffusion and thermodiffusion phenomena taking place in closed systems, within a certain delay after the beginning of the experiments it can always be assumed that the condition (2.118) has already been reached with good approximation. It can be said that a multi-component continuous and closed system attains the state corresponding to (2.118) sooner than the moment, from which the conductive transport processes become dominant. Of course, in each diffusional experiment (2.118) can only be an approximation, because owing to the different molecular masses of the components the condition $\dot{\boldsymbol{v}} = 0$ is not fulfilled exactly. However, the accelerations arising thus are, so small that they can be neglected and the same is true for pressure gradients brought about by this acceleration. Nevertheless, in absence of external forces and in the case of the fulfilment of (2.118), the pressure gradient is so small that it can be neglected at the beginning of each diffusional process.

7. Balances of Angular Momentum

The symmetry of the pressure tensor in more general cases is in a direct relation to the conservation of the angular momentum. For the sake of its analysis let us define the so-called "external" angular momentum of a unit mass moving with $\boldsymbol{v}$ barycentric velocity. It is, by definition

$$\boldsymbol{S}^{\mathrm{e}} \equiv \boldsymbol{r} \times \boldsymbol{v} \qquad (2.119)$$

with respect to the origin of a reference system determined by the Eulerian coordinates x_1, x_2, x_3. The substantial time derivative of this "external" (or often called mechanical) angular momentum is

$$\dot{\boldsymbol{S}}^{\mathrm{e}} = \boldsymbol{r} \times \dot{\boldsymbol{v}} \qquad (2.120)$$

since $\dot{\boldsymbol{r}} \times \boldsymbol{v}$ vanishes. The substantial balance of the mechanical angular momentum $\boldsymbol{S}^{\mathrm{e}}$ can immediately be obtained from (2.71) for the case $\boldsymbol{F} = 0$, i.e. when only the surface forces $\nabla \cdot \mathbf{P}$ act upon the continuum.

The vectorial product of (2.71) by $\boldsymbol{r}$ from the left, yields, that

$$\varrho\dot{\boldsymbol{S}}^{e} + \boldsymbol{r}\times\nabla\cdot\mathbf{P} = 0\,. \tag{2.121}$$

Comparing this with (2.15) it can be seen that (2.121) does not yet correspond to a general balance. Total agreement can be obtained, if the second term of (2.121) is transformed with (2.115) by also taking into consideration that

$$\mathbf{P} - \widetilde{\mathbf{P}} = 2\boldsymbol{P}^{va}, \tag{2.122}$$

where $\boldsymbol{P}^{va}$ the axial vector, which is created from the $\mathbf{P}^{va}$ antisymmetric part of the pressure tensor in the sense of (2.106). From (2.121) with (2.115) and (2.122) the balance

$$\varrho\dot{\boldsymbol{S}}^{e} + \nabla\cdot(\boldsymbol{r}\times\mathbf{P}) = 2\boldsymbol{P}^{va} \tag{2.123}$$

is obtained, which is the substantial balance of the mechanical angular momentum with the source term $2\boldsymbol{P}^{va}$. Since this was derived from (2.71) under the condition that external forces do not act upon the system, it is evident on the basis of (2.29), giving the classification of the source terms, that $2\boldsymbol{P}^{va}$ can represent only the "internal" source of the angular momentum. Hence

$$\boldsymbol{\sigma}^{i}_{S^e} = 2\boldsymbol{P}^{va} \tag{2.124}$$

is the "internal" source of the angular momentum and

$$\mathbf{J}_{S^e} = \boldsymbol{r}\times\mathbf{P} \tag{2.125}$$

is the substantial current density.

The local form balance equation of the mechanical angular momentum is immediately obtained, if the local current density $\mathbf{J}^{0}_{S^e}$ of the $\boldsymbol{S}^{e}$ angular momentum is introduced and by making use of (2.125) and (2.12) with $a \equiv S^{e}_{\alpha}$, $(\alpha = 1, 2, 3)$:

$$\mathbf{J}^{0}_{S^e} = \boldsymbol{r}\times\mathbf{P} + \varrho\boldsymbol{S}^{e}\boldsymbol{v}\,. \tag{2.126}$$

Here the dyad $\varrho\boldsymbol{S}^{e}\boldsymbol{v}$ is the convective current density of the angular momentum. From (2.123) by using (2.126) and (2.21) the local form of the balance of the angular momentum

$$\frac{\partial\varrho\boldsymbol{S}^{e}}{\partial t} + \nabla\cdot(\boldsymbol{r}\times\mathbf{P} + \varrho\boldsymbol{S}^{e}\boldsymbol{v}) = 2\boldsymbol{P}^{va} \tag{2.127}$$

is obtained, which is equivalent to (2.123).

It is shown by the balance equations of $\boldsymbol{S}^{e}$ that the source density interpreted in (2.124) vanishes if and only if the "external" or mechanical angular momentum is a conservative quantity, hence if

$$\mathbf{P} = \widetilde{\mathbf{P}}, \qquad \text{i.e.}\quad \mathbf{P}^{a} = 0 \tag{2.128}$$

is valid, that is, if the pressure tensor is symmetric. In other words, the symmetry of the pressure tensor is equivalent to the conservation of the mechanical ("external") angular momentum $\boldsymbol{S}^e$.

In the course of the development of thermodynamics of continuous systems up to recent times only little attention has been paid to the angular momentum. The reason for this was that no contribution to the entropy production is given by the balances of the angular momentum, when the investigations are restricted to the case of a symmetric pressure tensor. Recently, however, it has been demonstrated by GRAD [18], MEIXNER [19] and particularly by BARANOWSKI and ROMOTOWSKI [20] that the antisymmetric part of the pressure tensor is related to an "internal" angular momentum $\boldsymbol{S}^i$, by the change of which the source of the external mechanical angular momentum is compensated. The transformation of the "external" and "internal" angular momenta into each other is an irreversible process, by which an additional term is given to the entropy production (Chapter III). The treatment of such effects in hydrodynamics is possible with the aid of the so-called generalized Navier-Stokes equation, which is obtained by taking into account the antisymmetric part of the pressure tensor [3, 20]. In other cases (since in general $\boldsymbol{S}^i$ is the resultant of the angular momenta of atoms and molecules, respectively of the spins of nuclei and electrons per unit mass) by the transformations of "external" and "internal" angular momenta into one another, the treatment of further effects is also made possible with the aid of non-equilibrium thermodynamics as, for instance, the Einstein-de Haas effect [19]. Nevertheless it seems advisable to take into account more thoroughly the balances of angular momentum as it was hitherto done and therefore the above equations will be analysed further.

Let us define the "internal" angular momentum $\boldsymbol{S}^i$ which is assumed to be related to an internal rotation of the selected mass element of the continuum. Denoting the macroscopic mean of the inertia momentum of corpuscles forming the unit mass of the continuum by Θ and the average of the angular velocity of the internal rotation by $\boldsymbol{\omega}$, by definition

$$\boldsymbol{S}^i \equiv \Theta \boldsymbol{\omega} \tag{2.129}$$

is the specific "internal" angular momentum. This can vary only due to the variation of the "external" momentum $\boldsymbol{S}^e$, at least in the case, when no external forces act upon the continuum. In order to describe the transformation between the "external" and "internal" angular momentum with the aid of non-equilibrium thermodynamics, the axial vector

$$\boldsymbol{S} = \boldsymbol{S}^e + \boldsymbol{S}^i \tag{2.130}$$

giving the total angular momentum of the unit mass of the continuum must be introduced. For this vector it is assumed that the substantial balance expressing the conservation of the total angular momentum

$$\varrho \dot{\boldsymbol{S}} + \nabla \cdot (\boldsymbol{r} \times \mathbf{P} + \Pi) = 0 \tag{2.131}$$

is satisfied. Here, with the use of tensor Π the general case has been formulated where the spatial transport of the "internal" angular momentum must be taken into consideration. Hence, Π is the substantial current density of the "internal" angular momentum, i.e.

$$\mathbf{J}_S = \boldsymbol{r} \times \mathbf{P} + \Pi = \mathbf{J}_{S^e} + \Pi \tag{2.132}$$

is the substantial current density of the total angular momentum. Putting this relation in the form of coordinate notation

$$J_{\alpha\beta\gamma} = x_\alpha P_{\beta\gamma} - x_\gamma P_{\beta\alpha} + \pi_{\alpha\beta\gamma}, \qquad (\alpha, \beta, \gamma = 1, 2, 3) \tag{2.133}$$

it can be seen immediately that the tensor Π is a polar tensor of third order, and in the two extreme indices an antisymmetric one. (The divergence prescribed in (2.131) must be formed with respect to the β index.)

The balance equation valid for the "internal" angular momentum $\boldsymbol{S}^i$ can be obtained, if the balance (2.123) of the $\boldsymbol{S}^e$ "external" momentum is substracted from (2.131). Hence

$$\varrho \dot{\boldsymbol{S}}^e + \nabla \cdot \Pi = -2\boldsymbol{P}^{va} \tag{2.134}$$

is the substantial balance of the "internal" angular momentum with the "internal" source:

$$\boldsymbol{\sigma}^i_{Si} = -2\boldsymbol{P}^{va}. \tag{2.135}$$

It is shown by the given balances that in the case of symmetric $\mathbf{P}$, the "internal" source terms

$$\boldsymbol{\sigma}^i_{Se} = -\boldsymbol{\sigma}^i_{Si} = 2\boldsymbol{P}^{va} \tag{2.136}$$

vanish. Thus in such cases according to (2.123) and (1.234), respectively, the "external" and "internal" angular momenta are separate conservative quantities. In general, however, only the balance (2.131) expressing the conservation of the total angular momentum holds.

As a matter of course, if an external force $\boldsymbol{F}$ per unit mass acts upon mass elements of the continuum the balance of angular momentum

$$\varrho \dot{\boldsymbol{S}} + \nabla \cdot (\boldsymbol{r} \times \mathbf{P} + \Pi) = \varrho(\boldsymbol{r} \times \boldsymbol{F}) \tag{2.137}$$

must be employed instead of (2.131). In this case this equation expresses the conservation of the total angular momentum as well, since, in the sense of the general splitting of (2.29), the source term can be interpreted as an "external" source density

$$\sigma_{\mathbf{S}}^{\mathrm{e}} = \varrho(\boldsymbol{r} \times \boldsymbol{F}). \qquad (2.138)$$

The above-mentioned balances can be extended to the case of multi-component continua, if the pressure tensor $\mathbf{P}$ is assumed to be that of (2.85). A particular problem arises if the partial pressure tensors $\mathbf{P}_k$ of the component media are not symmetric. Hitherto such cases have not been examined, though it is evident that such examinations lead to the discovery of new effects, eventually to the development of new separation methods of components.

8. Balances of Kinetic Energy

Let us begin the determination of the balance equations for the kinetic energy of translation referred to the centre of mass motion. For the sake of generality let us start from the coordinate form of the impulse balance (2.87)

$$\varrho \dot{v}_\alpha + \frac{\partial P_{\beta\alpha}}{\partial x_\beta} = \sum_{k=1}^{K} \varrho_k F_{k\alpha}, \qquad (\alpha = 1, 2, 3) \qquad (2.139)$$

valid for multi-component continua. Multiplying it scalarly by the barycentric velocity and using the identity

$$\frac{\partial}{\partial x_\beta}(P_{\beta\alpha} v_\alpha) = v_\alpha \frac{\partial P_{\beta\alpha}}{\partial x_\beta} + P_{\beta\alpha} \frac{\partial v_\alpha}{\partial x_\beta}, \qquad (\alpha, \beta = 1, 2, 3) \qquad (2.140)$$

the substantial kinetic energy balance

$$\varrho \dot{\varepsilon}_{\mathrm{t}} + \nabla \cdot (\mathbf{P} \cdot \boldsymbol{v}) = \widetilde{\mathbf{P}} : \nabla \boldsymbol{v} + \sum_{k=1}^{K} \varrho_k \boldsymbol{F}_k \cdot \boldsymbol{v} \qquad (2.141)$$

is obtained. Here

$$\varepsilon_{\mathrm{t}} \equiv \frac{1}{2} \boldsymbol{v}^2 \qquad (2.142\mathrm{a})$$

is the specific translational barycentric kinetic energy and

$$\dot{\varepsilon}_{\mathrm{t}} = \boldsymbol{v} \cdot \dot{\boldsymbol{v}} \qquad (2.142\mathrm{b})$$

is its substantial time derivative. A comparison of (2.141) with the general (2.15) shows that

$$\boldsymbol{J}_{\varepsilon_{\mathrm{t}}} = \mathbf{P} \cdot \boldsymbol{v} \qquad (2.143)$$

is the substantial current density of the translational kinetic energy. Should we assume (as it is usual in ordinary hydrodynamics) that the pressure tensor is symmetric, the pressure tensor itself could be written instead of $\widetilde{\mathbf{P}}$ in the balance (2.141). However, the treatment of several effects would be excluded by this unnecessary restriction.

The local form balance of the kinetic energy of translation is similarly obtained from (2.74), or with the use of (2.139) and (2.21). In both ways we arrive at the local balance of the barycentric translational kinetic energy

$$\frac{\partial \varrho \varepsilon_t}{\partial t} + \nabla \cdot (\boldsymbol{J}_{\varepsilon_t} + \varrho \varepsilon_t \boldsymbol{v}) = \widetilde{\mathbf{P}} : \nabla \boldsymbol{v} + \sum_{k=1}^{K} \varrho_k \boldsymbol{F}_k \cdot \boldsymbol{v} \tag{2.144}$$

whose comparison with (2.8) shows that the local current density of the kinetic energy is

$$\boldsymbol{J}^0_{\varepsilon_t} = \boldsymbol{J}_{\varepsilon_t} + \varrho \varepsilon_t \boldsymbol{v} \tag{2.145}$$

where $\varrho \varepsilon_t \boldsymbol{v}$ is the current density of the kinetic energy of the convective motion.

In the given balances the total source density of the kinetic energy can be split up into two parts:

$$\sigma_{\varepsilon_t} = \sigma^{e}_{\varepsilon_t} + \sigma^{i}_{\varepsilon_t}, \tag{2.146}$$

i.e. for an "external" source density due to external forces

$$\sigma^{e}_{\varepsilon_t} = \sum_{k=1}^{K} \varrho_k \boldsymbol{F}_k \cdot \boldsymbol{v} = \varrho \boldsymbol{F} \cdot \boldsymbol{v} \tag{2.147}$$

and for an "internal" source density

$$\sigma^{i}_{\varepsilon_t} = \widetilde{\mathbf{P}} : \nabla \boldsymbol{v} \tag{2.148}$$

arising from the inhomogeneity which is represented by the tensor $\nabla \boldsymbol{v}$ in the internal of the system. Consequently, on the basis of what has been said with respect to (2.29) about the "external" and "internal" source densities, the kinetic energy of translation is a conservative quantity, if $\sigma^{i}_{\varepsilon_t} = 0$. It is shown by balances (2.141) and (2.144) that in each case, when internal inhomogeneities describable by the tensor $\nabla \boldsymbol{v}$ exist in the continuum, the translational kinetic energy is not conserved. In such cases the pressure tensor cannot be zero for any case of continuum models. Consequently, it is customary to speak of the dissipation of kinetic energy in case of $\sigma^{i}_{\varepsilon_t} \neq 0$. From a practical viewpoint this case is of particular importance, when the dissipation of the kinetic energy is determined by the viscous part $\mathbf{P}^{v}$ of the pressure tensor and by the velocity gradient $\nabla \boldsymbol{v}$. The Joule-experiment is a classical case for such dissipation of kinetic energy due to viscosity.

Let us split up the tensor $\nabla \boldsymbol{v}$ in order to analyse the "internal" source term of the kinetic energy of translation in more detail, as it has been done in the case of the viscous pressure tensor in (2.103) or in

(2.104). Accordingly, we have

$$\nabla v = \frac{1}{3}(\nabla \cdot v)\,\boldsymbol{\delta} + (\overset{0}{\nabla v})^{\mathrm{s}} + (\nabla v)^{\mathrm{a}} \tag{2.149}$$

where

$$\nabla \cdot v = \nabla v : \boldsymbol{\delta} = \frac{\partial v_\gamma}{\partial x_\gamma}, \qquad (\gamma = 1, 2, 3) \tag{2.150}$$

is the trace of the tensor ∇v, which is an invariant scalar and, according to (2.17) and (2.25), is equal to the deformation (dilatation or compression) of the continuum per unit volume and unit time. On the other hand, for an incompressible continuum

$$\nabla \cdot v = 0 \tag{2.151}$$

is valid. The meaning of the other terms is the following. The tensor ∇v can always be split into a symmetric

$$(\nabla v)^{\mathrm{s}}_{\alpha\beta} = \frac{1}{2}\left(\frac{\partial v_\beta}{\partial x_\alpha} + \frac{\partial v_\alpha}{\partial x_\beta}\right), \qquad (\alpha, \beta = 1, 2, 3) \tag{2.152}$$

and an antisymmetric part:

$$(\nabla v)^{\mathrm{a}}_{\alpha\beta} = \frac{1}{2}\left(\frac{\partial v_\beta}{\partial x_\alpha} - \frac{\partial v_\alpha}{\partial x_\beta}\right), \qquad (\alpha, \beta = 1, 2, 3) \tag{2.153}$$

i.e.

$$\nabla v = (\nabla v)^{\mathrm{s}} + (\nabla v)^{\mathrm{a}}, \tag{2.154}$$

and consequently, it is shown by (2.149) that

$$(\overset{0}{\nabla v})^{\mathrm{s}}_{\alpha\beta} = \frac{1}{2}\left(\frac{\partial v_\beta}{\partial x_\alpha} + \frac{\partial v_\alpha}{\partial x_\beta}\right) - \frac{1}{3}\,\delta_{\alpha\beta}\,\frac{\partial v_\gamma}{\partial x_\gamma}, \qquad (\alpha, \beta, \gamma = 1, 2, 3) \tag{2.155}$$

is a symmetric tensor with zero trace. This relation can also be symbolically given as

$$(\nabla v)^{\mathrm{s}} = (\overset{0}{\nabla v})^{\mathrm{s}} + \frac{1}{3}(\nabla \cdot v)\,\boldsymbol{\delta}. \tag{2.156}$$

On the other hand, from the components of the antisymmetric tensor given in (2.153), the axial vector wellknown in hydrodynamics as *vorticity* vector can be constructed:

$$\frac{1}{2}\nabla \times v = (\nabla v)^{\mathrm{a}}. \tag{2.157}$$

Let us form the scalar product of the viscous pressure tensor and the velocity gradient tensor with the use of (2.103) and (2.149). We have

$$\mathbf{P}^{\mathrm{v}} : \nabla v = p^{\mathrm{v}}(\nabla \cdot v) + \overset{0}{\mathbf{P}}{}^{\mathrm{vs}} : (\overset{0}{\nabla v})^{\mathrm{s}} + \mathbf{P}^{\mathrm{va}} : (\nabla v)^{\mathrm{a}} \tag{2.158}$$

because the scalar product of a symmetric and an antisymmetric tensor always vanishes. On the basis of this and (2.94), the "internal" source density (2.148) of the translation kinetic energy can be rewritten in the detailed form

$$\sigma^{\mathrm{i}}_{\varepsilon_t} = (p + p^{\mathrm{v}})\, \nabla \cdot \boldsymbol{v} + \overset{0}{\mathbf{P}}{}^{\mathrm{vs}} : \overset{0}{(\nabla \boldsymbol{v})}{}^{\mathrm{s}} - \mathbf{P}^{\mathrm{va}} : (\nabla \boldsymbol{v})^{\mathrm{a}} \qquad (2.159)$$

by taking into account the relations

$$\mathbf{P}^{\mathrm{vs}} = \widetilde{\mathbf{P}}^{\mathrm{vs}} \qquad (2.160\,\mathrm{a})$$

and

$$\mathbf{P}^{\mathrm{va}} = -\, \widetilde{\mathbf{P}}^{\mathrm{va}}. \qquad (2.160\,\mathrm{b})$$

It is often advantageous to write the scalar product of the antisymmetric tensor in the form of the scalar product of the axial vectors which can be formed from them. In this case the last term of (2.159) can be transformed with the use of (2.106) and (2.157) according to the identity

$$\mathbf{P}^{\mathrm{va}} : (\nabla \boldsymbol{v})^{\mathrm{a}} \equiv -\boldsymbol{P}^{\mathrm{va}} \cdot (\nabla \times \boldsymbol{v}), \qquad (2.161)$$

hence

$$\sigma^{\mathrm{i}}_{\varepsilon_t} = (p + p^{\mathrm{v}})\, \nabla \cdot \boldsymbol{v} + \overset{0}{\mathbf{P}}{}^{\mathrm{vs}} : \overset{0}{(\nabla \boldsymbol{v})}{}^{\mathrm{s}} + \boldsymbol{P}^{\mathrm{va}} \cdot (\nabla \times \boldsymbol{v}). \qquad (2.162)$$

The "internal" source density of the translational kinetic energy (2.162) refers to such isotropic continua, whose pressure tensor (owing to the non-zero antisymmetric part of the viscous pressure tensor) is not symmetric. Since, for such continua, according to (2.134), the angular momentum has an "internal" source term, which belongs to the internal rotation of the angular velocity $\boldsymbol{\omega}$, the existence of a rotational kinetic energy per unit mass can be assumed, to which an "internal" source density $\sigma^{\mathrm{i}}_{\varepsilon_r}$ belongs. Let us define the macroscopic specific kinetic energy of the internal rotation in the usual way

$$\varepsilon_{\mathrm{r}} \equiv \frac{1}{2} \Theta \omega^2 \qquad (2.163\,\mathrm{a})$$

the substantial time derivative of which is

$$\dot{\varepsilon}_{\mathrm{r}} = \Theta \boldsymbol{\omega} \cdot \dot{\boldsymbol{\omega}}. \qquad (2.163\,\mathrm{b})$$

For this quantity from (2.134) (neglecting the term containing the divergence for simplicity's sake) after the scalar multiplication by $\boldsymbol{\omega}$, we get the substantial balance

$$\varrho \dot{\varepsilon}_{\mathrm{r}} = \varrho \boldsymbol{\omega} \cdot \dot{\boldsymbol{S}}^{\mathrm{i}} = -\boldsymbol{\omega} \cdot 2\boldsymbol{P}^{\mathrm{va}}, \qquad (\equiv \boldsymbol{\omega} : \mathbf{P}^{\mathrm{va}}) \qquad (2.164)$$

with the internal production density

$$\sigma^{\mathrm{i}}_{\varepsilon_t} = -2\boldsymbol{\omega} \cdot \boldsymbol{P}^{\mathrm{va}}, \qquad (\equiv \boldsymbol{\omega} : \mathbf{P}^{\mathrm{va}}). \qquad (2.165)$$

Since in (2.162) the last term vanishes if the pressure tensor is symmetric and the same is also valid for the "internal" source density of the kinetic energy of rotation in (2.165), it is evident that the last source term of the translational kinetic energy is in a relation with the "internal" source of the rotational kinetic energy. This relation can easily be demonstrated.

Let us define the total specific kinetic energy by

$$\varepsilon_{\mathrm{k}} = \varepsilon_{\mathrm{t}} + \varepsilon_{\mathrm{r}} = \frac{1}{2}\boldsymbol{v}^2 + \frac{1}{2}\Theta\boldsymbol{\omega}^2. \tag{2.166}$$

For this quantity the substantial balances can immediately be given. By adding the balances (2.141) and (2.164) we have

$$\varrho\dot{\varepsilon}_{\mathrm{k}} + \nabla \cdot (\mathbf{P} \cdot \boldsymbol{v}) = \sigma^{\mathrm{e}}_{\varepsilon_{\mathrm{k}}} + \sigma^{\mathrm{i}}_{\varepsilon_{\mathrm{k}}}, \tag{2.167}$$

where

$$\sigma^{\mathrm{i}}_{\varepsilon_{\mathrm{k}}} = (p + p^{\mathrm{v}})\, \nabla \cdot \boldsymbol{v} + \overset{0}{\mathbf{P}}{}^{\mathrm{vs}} : \overset{0}{(\nabla\boldsymbol{v})}{}^{\mathrm{s}} + \boldsymbol{P}^{\mathrm{va}} \cdot (\nabla\times\boldsymbol{v} - 2\boldsymbol{\omega}) \tag{2.168}$$

is the "internal" source density. Here the last term represents a kinetic energy source, which is produced as a result of the transformation of the kinetic energy of translational motion and the kinetic energy of the internal rotation into one another. This term (with the exception of the very particular case, when the vectors $\boldsymbol{P}^{\mathrm{va}}$ and $(\nabla\times\boldsymbol{v} - 2\boldsymbol{\omega})$ are normal to each other) becomes zero in two essentially different cases. First, in the case of the symmetric pressure tensor ($\mathbf{P}^{\mathrm{a}} = \mathbf{P}^{\mathrm{va}} = 0$), already examined in detail, hence when the mass elements of the continuum do not rotate. The second case is, when

$$\boldsymbol{\omega} = \frac{1}{2}\nabla\times\boldsymbol{v} \tag{2.169}$$

hence, the vorticity vector $\frac{1}{2}\nabla\times\boldsymbol{v}$ is just equal to the angular velocity $\boldsymbol{\omega}$, determining the internal "rigid like" rotation of the mass elements of the continuum. Assuming that the vorticity vector is uniform and constant in space and zero at the beginning, it can be confirmed that the equality (2.169) is valid for most of the fluids after a very short time of relaxation. After this equalization the last source term in (2.168) need not be taken into account. As will be seen in the following, in such cases the "internal" source (2.168) of the kinetic energy is characteristic for a continuum, whose dynamic behavior is governed by the Navier-Stokes equation corresponding to the symmetric pressure tensor. In other cases, when the equality (2.169) does not hold, an additive term, describing the rotational viscosity, is added to the classical Navier-Stokes equations (see Chapter VI).

Before examining a multi-component hydrodynamics continuum, some complementary remarks must be made. The balance (2.141) and (2.144) are evidently valid also for the case of multi-component continua. However, it is important that it is not the total translational kinetic energy

$$\varepsilon_t^t \equiv \frac{1}{2}\varrho^{-1}\sum_{k=1}^{K}\varrho_k \boldsymbol{v}_k^2 = \frac{1}{2}\sum_{k=1}^{K} c_k \boldsymbol{v}_k^2 \tag{2.170}$$

but only the barycentric kinetic energy $\varepsilon_t \equiv \frac{1}{2}\boldsymbol{v}^2$ which occurs in the balances given for such continua. The latter can be transformed with the aid of (1.39), and from the relation obtained

$$\varepsilon_t \equiv \frac{1}{2}\boldsymbol{v}^2 = \frac{1}{2}\sum_{k=1}^{K} c_k \boldsymbol{v}_k^2 - \frac{1}{2}\sum_{k=1}^{K} c_k(\boldsymbol{v}_k - \boldsymbol{v})^2 \tag{2.171}$$

it is apparent that the total kinetic energy of translation is determined by

$$\varepsilon_t^t = \varepsilon_t + \frac{1}{2}\sum_{k=1}^{K} c_k(\boldsymbol{v}_k - \boldsymbol{v})^2 . \tag{2.172}$$

The last term can be given owing to (1.41) also as

$$\varepsilon_d \equiv \frac{1}{2}\sum_{k=1}^{K} c_k \boldsymbol{w}_k^2 , \tag{2.173}$$

from which it can be seen that ε_d is the kinetic energy of the components corresponding to their diffusion relative to the local barycenter. It is shown by (2.171) that with the formulation of each relation given for the barycentric kinetic energy $\varepsilon_t \equiv \frac{1}{2}\boldsymbol{v}^2$, the kinetic energy of diffusion (2.173) is a priori substracted from the total translational kinetic energy (2.170). There is not doubt that the kinetic energy of diffusion can in the majority of cases be neglected with respect to the barycentric kinetic energy, however, at least in principle all what has been said before must be taken into account. It will be seen in the following that in the case of multi-component systems the foregoings contribute to the definition of the specific internal energy.

9. Balances of Potential Energy

Now let our examinations be confined to external fields which are conservative and with respect to the unit mass of the components can be characterized by some scalar potential φ_k. Accordingly, let

$$\boldsymbol{F}_k = -\nabla\varphi_k \tag{2.174a}$$

and

$$\frac{\partial\varphi_k}{\partial t} = 0 \tag{2.174b}$$

be valid and let us study the balance equation of the potential energy

$$\varphi = \sum_{k=1}^{K} c_k \varphi_k \tag{2.175a}$$

or

$$\varrho\varphi = \sum_{k=1}^{K} \varrho_k \varphi_k . \tag{2.175b}$$

First of all let us examine under what condition (2.174a) is compatible with condition (2.84) stipulated for the specific forces, further, with the derivation

$$\boldsymbol{F} = -\nabla\varphi \tag{2.176}$$

of the resultant force $\boldsymbol{F}$ from a scalar potential φ. It follows from (2.84), (2.174a) and (2.176) that

$$\nabla\varphi = \sum_{k=1}^{K} c_k \nabla\varphi_k \tag{2.177}$$

is valid. Since it is evident that for φ the condition

$$\frac{\partial^2\varphi}{\partial x_\alpha\, \partial x_\beta} = \frac{\partial^2\varphi}{\partial x_\beta\, \partial x_\alpha}, \qquad (\alpha, \beta = 1, 2, 3) \tag{2.178}$$

is valid, the necessary and sufficient condition for the relations (2.174a), (2.84) and (2.176) to be compatible with one another is the following condition

$$\sum_{k=1}^{K} \left[\frac{\partial c_k}{\partial x_\alpha}\frac{\partial \varphi_k}{\partial x_\beta} - \frac{\partial c_k}{\partial x_\beta}\frac{\partial \varphi_k}{\partial x_\alpha}\right] = \sum_{k=1}^{K} \frac{\partial(c_k, \varphi_k)}{\partial(x_\alpha, x_\beta)} = 0 . \tag{2.179}$$

As can easily be demonstrated, this condition is satisfied in case of gravitational and electrostatic fields. In the latter case, of course, only with the additional stipulation that the specific charges e_k of the components are constant.

Let us determine the balance equations of the potential energy φ. The multiplication of the local component balances (2.31) by φ_k and after summing over the components, the local balance

$$\frac{\partial\varrho\varphi}{\partial t} + \nabla\cdot \sum_{k=1}^{K} \varphi_k \boldsymbol{J}_k^0 = \sum_{k=1}^{K} \boldsymbol{J}_k^0 \cdot \boldsymbol{F}_k + \sum_{k=1}^{K} \sum_{j=1}^{R} \varphi_k \nu_{kj} J_j \tag{2.180}$$

is obtained, with application of relations (2.44), (2.174a) and (2.175). Eliminating the component current densities $\boldsymbol{J}_k^0$ with (1.37) and (1.42), the balance

$$\begin{aligned} &\frac{\partial\varrho\varphi}{\partial t} + \nabla\cdot\left(\sum_{k=1}^{K} \varphi_k \boldsymbol{J}_k + \varrho\varphi\boldsymbol{v}\right) \\ &= -\sum_{k=1}^{K} \varrho_k \boldsymbol{F}_k \cdot \boldsymbol{v} - \sum_{k=1}^{K} \boldsymbol{J}_k \cdot \boldsymbol{F}_k + \sum_{k=1}^{K} \sum_{j=1}^{R} \varphi_k \nu_{kj} J_j \end{aligned} \tag{2.181}$$

is obtained in terms of the diffusion current densities $\boldsymbol{J}_k$. It is shown by comparing this balance equation with (2.8) that the local current density of the potential energy is

$$\boldsymbol{J}_\varphi^0 = -\sum_{k=1}^{K} \varphi_k \boldsymbol{J}_k + \varrho\varphi\boldsymbol{v} \tag{2.182}$$

where $\varrho\varphi\boldsymbol{v}$ is the convective term. From the comparison it becomes clear in the sense of (2.29) that the first two terms on the right-hand side of (2.181) should be interpreted as "external" source terms, whereas the double sum in the last term should be considered as an "internal" source term of the potential energy. Accordingly,

$$\sigma_\varphi^{\mathrm{e}} = -\sum_{k=1}^{K} \varrho_k \boldsymbol{F}_k \cdot \boldsymbol{v} - \sum_{k=1}^{K} \boldsymbol{J}_k \cdot \boldsymbol{F}_k = -\sum_{k=1}^{K} \varrho_k \boldsymbol{F}_k \cdot \boldsymbol{v}_k \tag{2.183}$$

is the "external", whereas

$$\sigma_\varphi^{\mathrm{i}} = \sum_{k=1}^{K} \sum_{j=1}^{R} \varphi_k \nu_{kj} J_j \tag{2.184}$$

is the "internal" source density of the potential energy. It is remarkable that the "internal" source density vanishes in each case, when in the course of the chemical reaction the condition

$$\sum_{k=1}^{K} \varphi_k \nu_{kj} = 0, \qquad (j = 1, 2, \ldots, R) \tag{2.185}$$

is fulfilled. The fulfilment of this condition is ensured in the case of a gravitational field by the conservation of mass, whereas in the case of an electrostatic field by the conservation of charge.

The substantial balance of the potential energy can also be obtained readily. Namely, (2.181) leads, by applying (2.21) for the case $a \equiv \varphi$, directly to the substantial balance equation:

$$\varrho\dot{\varphi} + \nabla \cdot \sum_{k=1}^{K} \varphi_k \boldsymbol{J}_k = \sigma_\varphi^{\mathrm{e}} + \sigma_\varphi^{\mathrm{i}}. \tag{2.186}$$

This corresponds to the general (2.15), if, by definition

$$\boldsymbol{J}_\varphi = \sum_{k=1}^{K} \varphi_k \boldsymbol{J}_k = \sum_{k=1}^{K} \varphi_k \boldsymbol{J}_k^0 - \varrho\varphi\boldsymbol{v} \tag{2.187}$$

is the substantial current density of the potential energy, further, if $\sigma_\varphi^{\mathrm{e}}$ and $\sigma_\varphi^{\mathrm{i}}$ are given by (2.183) and (2.184).

10. Balances of Mechanical Energy

Let us define the specific mechanical energy as the sum of the translational kinetic energy ε_{t}, the rotational kinetic energy ε_{r}, and the potential energy φ. Hence

$$\varepsilon_{\mathrm{m}} \equiv \frac{1}{2}\boldsymbol{v}^2 + \frac{1}{2}\Theta\boldsymbol{\omega}^2 + \varphi \tag{2.188}$$

is the specific mechanical energy. Concerning this quantity two remarks must be made immediately. Firstly ε_{m} is only equal to the total mechanical energy of the unit mass when the continuum has a single component. In case of multi-component continua, the total mechanical energy must be given by the total translational kinetic energy introduced in (2.170) as will be done later on. Secondly, because the specific potential energy φ in (2.188) may not only be a mechanical energy, but, for instance, can have an electrical character as well and thus for ε_{m} the usually applied denomination "mechanical energy" is not quite appropriate.

By adding (2.167) and (2.186) we obtain the following substantial balances

$$\varrho\dot{\varepsilon}_{\mathrm{m}} + \nabla\cdot\boldsymbol{J}_{\varepsilon_{\mathrm{m}}} = \sigma_{\varepsilon_{\mathrm{m}}} \tag{2.189}$$

where

$$\boldsymbol{J}_{\varepsilon_{\mathrm{m}}} = \sum_{k=1}^{K}\varphi_k\boldsymbol{J}_k + \mathbf{P}\cdot\boldsymbol{v} \tag{2.190}$$

is the substantial energy current density and

$$\begin{aligned}\sigma_{\varepsilon_{\mathrm{m}}} &= \sigma_{\varepsilon_{\mathrm{t}}} + \sigma_{\varepsilon_{\mathrm{r}}} + \sigma_{\varepsilon_{\varphi}} = \sigma^{\mathrm{i}}_{\varepsilon_{\mathrm{t}}} + \sigma^{\mathrm{e}}_{\varepsilon_{\mathrm{t}}} + \sigma^{\mathrm{i}}_{\varepsilon_{\mathrm{r}}} + \sigma^{\mathrm{i}}_{\varepsilon_{\varphi}} + \sigma^{\mathrm{e}}_{\varepsilon_{\varphi}} \\ &= (p + p^{\mathrm{v}})\,\nabla\cdot\boldsymbol{v} + \overset{0}{\mathbf{P}}{}^{\mathrm{vs}} : \overset{0}{(\nabla\boldsymbol{v})}{}^{\mathrm{s}} + \boldsymbol{P}^{\mathrm{va}}\cdot(\nabla\times\boldsymbol{v} - 2\boldsymbol{\omega}) \\ &\quad - \sum_{k=1}^{K}\boldsymbol{J}_k\cdot\boldsymbol{F}_k + \sum_{k=1}^{K}\sum_{j=1}^{R}\varphi_k\nu_{kj}J_j\end{aligned} \tag{2.191}$$

is the source density of the mechanical energy. Expressions (2.186), (2.183) and (2.184) have also been used for the derivation of this detailed form and thus it follows from their meaning that in (2.191) the quantity

$$\begin{aligned}\sigma^{\mathrm{i}}_{\varepsilon_{\mathrm{m}}} &= (p + p^{\mathrm{v}})\,\nabla\cdot\boldsymbol{v} + \overset{0}{\mathbf{P}}{}^{\mathrm{vs}} : \overset{0}{(\nabla\boldsymbol{v})}{}^{\mathrm{s}} + \boldsymbol{P}^{\mathrm{va}}\cdot(\nabla\times\boldsymbol{v} - 2\boldsymbol{\omega}) \\ &\quad + \sum_{k=1}^{K}\sum_{j=1}^{R}\varphi_k\nu_{kj}J_j\end{aligned} \tag{2.192}$$

should be interpreted as the "internal" source density of the mechanical energy. It is only the quantity

$$\sigma^{\mathrm{e}}_{\varepsilon_{\mathrm{m}}} = \sigma^{\mathrm{e}}_{\varepsilon_{\mathrm{t}}} + \sigma^{\mathrm{e}}_{\varphi} = -\sum_{k=1}^{K}\boldsymbol{J}_k\cdot\boldsymbol{F}_k \tag{2.193}$$

which is to be considered the "external" source density.

The local balance of the mechanical energy can readily be given with the use of local balances (2.144) and (2.181):

$$\frac{\partial \varrho \varepsilon_{\mathrm{m}}}{\partial t} + \nabla \cdot \boldsymbol{J}^0_{\varepsilon_{\mathrm{m}}} = \sigma_{\varepsilon_{\mathrm{m}}}, \tag{2.194}$$

where

$$\boldsymbol{J}^0_{\varepsilon_{\mathrm{m}}} = \sum_{k=1}^{K} \varphi_k \boldsymbol{J}_k + \mathbf{P} \cdot \boldsymbol{v} + \varrho \varepsilon_{\mathrm{m}} \boldsymbol{v} \tag{2.195}$$

is the local energy current density with the convective term $\varrho \varepsilon_{\mathrm{m}} \boldsymbol{v}$.

It has already been mentioned that in case of a multi-component continuum the specific mechanical energy ε_{m} defined in (2.188) does not yield the total specific mechanical energy, since ε_{m} includes only the barycentric kinetic energy ε_{t} instead of the total translational kinetic energy $\varepsilon^{\mathrm{t}}_{\mathrm{t}}$ defined in (2.172). Consequently, in case of multi-component continua, the total specific mechanical energy must be given by the quantity

$$\varepsilon^{\mathrm{t}}_{\mathrm{m}} \equiv \varepsilon_{\mathrm{t}} + \varepsilon_{\mathrm{d}} + \varepsilon_{\mathrm{r}} + \varepsilon_{\varphi} = \frac{1}{2} \boldsymbol{v}^2 + \frac{1}{2} \sum_{k=1}^{K} c_k \boldsymbol{w}_k^2 + \frac{1}{2} \Theta \boldsymbol{\omega}^2 + \varphi \tag{2.196}$$

according to (2.172), and this quantity is related to

$$\varepsilon_{\mathrm{m}} = \varepsilon^{\mathrm{t}}_{\mathrm{m}} - \varepsilon_{\mathrm{d}} \tag{2.197}$$

defined in (2.188). Thus in each balance equation which is valid for ε_{m}, the kinetic energy of diffusion is a priori subtracted from the total mechanical energy. This fact must be taken into account in each case, when there is a question of balances valid for energy quantities which contain only the barycentric kinetic energy $\varepsilon_{\mathrm{t}} \equiv \frac{1}{2} \boldsymbol{v}^2$ instead of the total translational kinetic energy $\varepsilon^{\mathrm{t}}_{\mathrm{t}}$.

Finally let us define the local current density of the total mechanical energy similarly to the current density $\boldsymbol{J}^0_{\varepsilon_{\mathrm{m}}}$ given in (2.195). This evidently is

$$\boldsymbol{J}^0_{\varepsilon^{\mathrm{t}}_{\mathrm{m}}} = \sum_{k=1}^{K} \varphi_k \boldsymbol{J}_k + \mathbf{P} \cdot \boldsymbol{v} + \varrho \varepsilon^{\mathrm{t}}_{\mathrm{m}} \boldsymbol{v}, \tag{2.198}$$

where the first sum gives the current density of the potential energy transferred by all the components, $\mathbf{P} \cdot \boldsymbol{v}$ is an energy flux due to the mechanical work performed on the system by surface forces, and finally $\varrho \varepsilon^{\mathrm{t}}_{\mathrm{m}} \boldsymbol{v}$ is a convection term related to the convection of the total energy.

In the above discussion only conservative external fields have been examined. However, the forces $\boldsymbol{F}_k$ may be mechanical and electromagnetic ones of any kind, and thus it is evident that also the effect of external fields of most diverse nature can be taken into consideration in the basic equations of the field theory. This is a direct way to complete the basic equations of continuum mechanics by the basic equations of non-conservative (electromagnetic) external fields. If the set of basic equations thus obtained is completed by relations which can be derived from the first and second law of thermodynamics, a field theory may be developed which includes all three classical disciplines of physics in an organic unity and is valid for a large class of phenomena. Even now it is possible to develop a unified field theory of continua, which is suitable for the simultaneous description of mechanical, electromagnetic and thermal state variations. We note that for example the development of thermo-electrodynamics of polarizable media is making advances [3, 21].

III. Thermodynamics of Continua

In the classical work by ONSAGER in 1931 [22] the exact fundamentals of the thermodynamic theory of irreversible processes were laid down already. Because of the summarizing works of PRIGOGINE [17, 23], MEIXNER [4, 24], DENBIGH [25], HAASE [26, 27], GYARMATI [28], FITTS [13], GUMINSKI [29], RYSSELBERGHE [30], and particularly by de Groot's monographs [3,8] at present we have a well-founded theory which at the same time is applicable to solutions of practical problems. While describing this theory (which is often called irreversible thermodynamics, non-equilibrium thermodynamics or Onsager's theory) in terms of the concepts of field theory, the first task is the reformulation of the fundamental laws of classical thermodynamics (reversible or equilibrium theory the most appropriate name of which is thermostatics) in local forms. For such a formulation—it has to be assumed—beyond the postulates of classical field theory only that for the volume elements (cells) of a continuum the hypothesis of local (cellular) equilibrium is valid. Accordingly, after the local formulation of the first and second law, the hypothesis of local equilibrium, further its validity limits for the case of non-equilibrium systems are dealt with. After the treatment of the basic problems, the total and general development of the theory is given, at least with respect to the models of multicomponent hydro-thermodynamic systems. The balance equations valid for the internal energy and thereafter the actual forms of the entropy balance, playing a central role in thermodynamics, are determined. This is followed by the deduction of the total linear Onsager theory together with the description of the Curie principle, and the determination of Onsager's reciprocal relations.

1. The Local Forms of the First and Second Law

a) The First law. Let us consider a system which is in a conservative external field and which is isolated from its surroundings with respect to each type of short-range interaction. A first integral of the equations of motion referring to the macroscopic coordinates and impulses of such a system is given by the energy integral

$$E_{\mathrm{m}} = E_{\mathrm{k}} + E_{\mathrm{p}} = \text{const.} \tag{3.1}$$

where E_{k} and E_{p} are the macroscopic kinetic and potential energy of the system. The constant quantity is the total mechanical energy for which the conservation law (3.1) or its differential form

$$\mathrm{d}E_{\mathrm{m}} = 0 \tag{3.2}$$

is valid.

Equations (3.1) and (3.2) give the conservation law of energy only in a macroscopical sense. However, it is shown by the examination of thermal phenomena, the molecular structure of bodies and by the dynamism of this structure that each body, apart from its mechanical energy determined by the macroscopic coordinates and impulses, possesses an additional energy content. Consequently, the energy E_{m} of (3.1) is complemented in thermodynamics by an energy function U, which is determined only by the internal microstate of the systems. This U function is called *internal energy*. According to the foregoings, the total energy content of a system from macroscopic and microscopic point of view is

$$E^{\mathrm{t}} = E_{\mathrm{m}} + U \tag{3.3}$$

where the function U is determined (at least strictly speaking) by a very large number of microparameters. Statistical physics is concerned with the determination of the dependence of the internal energy on the microparameters of corpuscles which constitute the system. In phenomenological thermodynamics equation (3.3) for the function U is a definition only, whose correctness is ensured by the general conservation law of energy. If a system is considered which is in an interaction with its surroundings, the conservation of the total energy is expressed in differential form by

$$\mathrm{d}E^{\mathrm{t}} = \mathrm{d}E_{\mathrm{m}} + \mathrm{d}U = 0\,. \tag{3.4}$$

According to the first law of thermodynamics for the internal energy variation $\mathrm{d}U$ the relation

$$\mathrm{d}U = đQ + đW \tag{3.5}$$

is valid[1] in the case of simultaneous heat effect and performance of work. Here $đW$ is the elementary work exerted by all the forces acting upon the system, whereas $đQ$ is the elementary heat-exchange due to the thermal interaction of the system with its surroundings, i.e. $đQ$ is the elementary heat quantity received ($đQ > 0$) or lost ($đQ < 0$) by the system.

The first law is expressed by (3.5) in a global form for an elementary state variation of the system. It will be seen that with the development of thermodynamics of continua, the local

$$\frac{\partial \varrho u}{\partial t} + \nabla \cdot \boldsymbol{J}_u^0 = \sigma_u \tag{3.6}$$

and the substantial

$$\varrho \dot{u} + \nabla \cdot \boldsymbol{J}_u = \sigma_u \tag{3.7}$$

balance equations of the specific internal energy u (which are immediately obtained from general balances (2.8) and (2.15) with the substitution $a \equiv u$) have to be established in such a way that they should be compatible with the equation

$$\mathrm{d}u = \mathrm{d}q + \mathrm{d}w \tag{3.8}$$

which is similar to global equation (3.5), but refers to specific quantities. The actual forms of balance equations (3.6) and (3.7) will be derived later on.

b) The Second Law. For the establishment of entropy balances, which play a central role in non-equilibrium thermodynamics, we have to start from

$$\mathrm{d}S = \mathrm{d_r}S + \mathrm{d_i}S \tag{3.9}$$

which determines the elementary entropy variation in general and has already been discovered by Clausius. In this global equation $\mathrm{d}S$ is the total entropy variation of the system. This quantity is composed of the external (or reversible) entropy variation

$$\mathrm{d_r}S = \frac{\mathrm{d_r}Q}{T} \tag{3.10}$$

belonging to the heat $\mathrm{d_r}Q$ exchanged in a reversible way with the surroundings of the system on the absolute temperature T, and of the positive entropy variation

$$\mathrm{d_i}S \geqq 0 \tag{3.11}$$

[1] Here operator $đ$ refers to the fact that, in general, the elementary energy variations (heat, work etc.) are not total differentials but Pfaffian forms of the equilibrium state parameters. Though, this fact is important and the above used denotation is common in the equilibrium theory in the followings it will be omitted.

which is due to irreversible processes taking place inside the system and which is zero in the reversible or equilibrium case. The above relations refer to reversible processes, if (3.11) expresses the case of equality. In the case of irreversible processes the given relations are valid in general and express the Carnot-Clausius theorem, i.e. the second (or entropy) law.

With the development of thermodynamics for continua the global relation (3.9) must be formulated in local form, and this local form leads immediately to the entropy balance required. Let us consider a continuum whose density and specific entropy is ϱ and s, respectively. In this case, to expression

$$\frac{\mathrm{d}S}{\mathrm{d}t} = \frac{\mathrm{d_r}S}{\mathrm{d}t} + \frac{\mathrm{d_i}S}{\mathrm{d}t} \tag{3.12}$$

which follows from (3.9) we can assign in addition to expression

$$S = \int_V \varrho s \, \mathrm{d}V \tag{3.13}$$

describing the total entropy of the system, also expressions

$$\frac{\mathrm{d_r}S}{\mathrm{d}t} = - \oint_\Omega \boldsymbol{J}_\mathrm{s} \cdot \mathrm{d}\boldsymbol{\Omega} \tag{3.14a}$$

and

$$\frac{\mathrm{d_i}S}{\mathrm{d}t} = \int_V \sigma \, \mathrm{d}V \geqq 0 . \tag{3.14b}$$

Here $\boldsymbol{J}_\mathrm{s}$ is the substantial entropy current density, and σ is the entropy production per unit volume and unit time, which is, according to (3.11), a definite positive quantity for irreversible transformations.

From the general balance equations (2.8) and (2.15), with the choice $a \equiv s$, as well as directly from relations (3.13) and (3.14) we arrive at the conclusion that (3.12) is a global entropy balance, and consequently in an arbitrary internal point of the continua the local

$$\frac{\partial \varrho s}{\partial t} + \nabla \cdot \boldsymbol{J}_\mathrm{s}^0 = \sigma \geqq 0 \tag{3.15}$$

or the substantial

$$\varrho \dot{s} + \nabla \cdot \boldsymbol{J}_\mathrm{s} = \sigma \geqq 0 \tag{3.16}$$

balances correspond to it. The difference between the local $\boldsymbol{J}_\mathrm{s}^0$ and the substantial entropy current density $\boldsymbol{J}_\mathrm{s}$ is, according to (2.12),

$$\boldsymbol{J}_\mathrm{s} = \boldsymbol{J}_\mathrm{s}^0 - \varrho s \boldsymbol{v} \tag{3.17}$$

where $\varrho s \boldsymbol{v}$ is the convective entropy current. From the method followed it is evident that entropy balances (3.15) and (3.16) are the local equivalences of the global balance (3.12) including also the second law with a source term σ which satisfies the inequality: $\sigma \geqq 0\cdot$ The actual expressions of the above entropy balances are derived later on.

c) The Condition of the Cellular (local) Equilibrium. The central problem of thermodynamics is the establishment of the entropy balances for actual system-models. However, actual entropy balances are obtained only if the balances derived for the fundamental mechanical and electromagnetic quantities are brought into relation with the balances referring to the thermodynamic quantities (internal energy, entropy, etc.).

Exact relations between the thermal and non-thermal variables can be obtained from the classical (thermostatic) theory. The relations in question are, however, unfortunately strictly valid only for equilibrium systems according to the spirit of the classical theory and within its validity range. Moreover, according to the classical heat theory, such quantities as temperature, entropy, ect. are not defined for non-equilibrium states. This difficulty is eliminated by the hypothesis of cellular (local) equilibrium. This gives a possibility to apply the equilibrium state parameters of thermostatics and the relations existing between them in the case of non-equilibrium systems as well. The cellular equilibrium of continuous systems involves the hypothesis that the continuum is the sum of elementary cells, in which equilibrium conditions are fulfilled, with good approximation, in spite of the fact that the total system represents irreversible and not equilibrium processes.

In a mathematical sense, the question is whether the thermostatic state description, valid for equilibrium systems, is valid for any elementary cell of the continuum. It is assumed that in the case of cellular equilibrium the Gibbs relation including both the first and the second laws for equilibrium systems can be applied in the usual form with respect to the unit mass of the continuum, i.e.

$$\mathrm{d}u = T\,\mathrm{d}s - p\,\mathrm{d}v + \sum_{k=1}^{K} \mu_k \mathrm{d}c_k \tag{3.18}$$

where μ_k is the chemical potential of the k-th component. The substantial time derivative of this relation given in the alternative form

$$\dot{s} = \frac{1}{T}\dot{u} + \frac{p}{T}\dot{v} - \sum_{k=1}^{K} \frac{\mu_k}{T}\dot{c}_k \tag{3.19}$$

does not mean any further restriction. Here the space and time dependence of the specific entropy cannot be considered as explicitly determined, but only through the space and time dependence of the variables $u = u(\boldsymbol{r}, t)$, $v = v(\boldsymbol{r}, t)$ and $c_k = c_k(\boldsymbol{r}, t)$, i.e. in an indirect way. Since the variation of the latter is given by the corresponding balance equations, it is evident that a possibility is given to determine the actual forms of the entropy balances by the balances in question and by relation (3.19).

From the Gibbs relation, (3.18), which is valid only for multi-component hydro-thermodynamic systems only the entropy balance of such systems can be determined. If we want to determine the entropy balance equations of other, more general system-models, (3.18) has to be modified and completed. This is the case for thermomechanical system-models of different elastic and plastic materials [16], in case of electromagnetic phenomena in polarized media [3, 21, 27], etc. All these cases can be summarized by the generalized Gibbs relation

$$\mathrm{d}u = \sum_{i=1}^{f} \Gamma_i^* \, \mathrm{d}a_i \tag{3.20}$$

where Γ_i^* is the thermostatic intensive quantity conjugated to the elementary reversible variation of the specific parameter a_i (for example in (3.18): T, $-p$, μ_k etc.). The fact that the different elementary reversible specific energy variations $\mathrm{d}w_i$ can always be written in the form $\mathrm{d}w_i = \Gamma_i^* \, \mathrm{d}a_i$ has to be proved, in general cases, just as the form $\mathrm{d}_r Q = T \, \mathrm{d}S$ (or $\mathrm{d}q = T \, \mathrm{d}s$) of the elementary reversible heat effect can be given on the basis of Caratheodory's axioms. The generalization of the exact Caratheodory's axiomatic development of thermostatics in the direction that (3.20) is valid in any reversible energy variation is self-explanatory [31, 32]. Consequently, (3.20) is called the energy picture (or representation) of the generalized Gibbs relation. The corresponding entropy picture

$$\mathrm{d}s = \sum_{i=1}^{f} \Gamma_i \mathrm{d}a_i \tag{3.21}$$

is obtained, if by chosing $\Gamma_1^* \equiv T$ and $a_1 \equiv s$ in (3.20)

$$\mathrm{d}u = T \, \mathrm{d}s + \sum_{i=2}^{f} \Gamma_i \, \mathrm{d}a_i, \tag{3.22}$$

whereby we have the expression

$$\mathrm{d}s = \frac{1}{T} \mathrm{d}u - \sum_{i=2}^{f} \frac{\Gamma_i^*}{T} \mathrm{d}a_i \tag{3.23}$$

for $\mathrm{d}s$, and, if the coefficients of the quantities $\mathrm{d}a_i$ are identified in (3.21) and (3.23), we derive:

$$\Gamma_1 \equiv \frac{1}{T}, \quad \Gamma_i \equiv -\frac{\Gamma_i^*}{T}, \qquad (i = 2, 3, \ldots, f). \tag{3.24}$$

Here the quantities Γ_i, $(i = 1, 2, \ldots, f)$ are the intensity quantities belonging to the entropy picture. (For example, in (3.19): $\frac{1}{T}, \frac{p}{T}, -\frac{\mu_k}{T}$, $(k = 1, 2, \ldots, K)$.) From the substantial derivative of the generalized Gibbs relation (3.21), i.e. from

$$\dot{s} = \sum_{i=1}^{f} \Gamma_i \dot{a}_i \tag{3.25}$$

the entropy balance equations can be determined in each case, when the substantial balances (2.15) are known for the quantities a_i. In the following we confine ourselves to the determination of the entropy balance equation valid in the case of multi-component hydro-thermodynamic systems only, sufficient basis to do so supplied by (3.18) and (3.19); however the generalizations given in equations (3.20) to (3.25), considering that also (3.18) and (3.19) are included in them, will prove to be useful in the following.

The postulate of local equilibrium for each cell of a continuum does not mean only the assumption of the validity of the Gibbs relation, but of all other thermostatic relations too. Thus, for example we accepted also the existence of the specific Gibbs function

$$g = \sum_{k=1}^{K} \mu_k c_k \tag{3.26}$$

and the wellknown relation

$$\mathrm{d}g = -s\,\mathrm{d}T + v\,\mathrm{d}p + \sum_{k=1}^{K} \mu_k\,\mathrm{d}c_k \tag{3.27}$$

from which with (3.18) we can derive the Gibbs-Duhem relation

$$\sum_{k=1}^{K} c_k\,\mathrm{d}\mu_k = -s\,\mathrm{d}T + v\,\mathrm{d}p \tag{3.28a}$$

or alternatively

$$\sum_{k=1}^{K} \varrho_k \mathrm{d}\mu_k = -\varrho s\,\mathrm{d}T + dp. \tag{3.28b}$$

This relation has also a fundamental importance in thermodynamics of multi-component systems, and can be given (expressing the spatial variations of the variables μ_k, T and p) in the form:

$$\sum_{k=1}^{K} \varrho_k\,\nabla\mu_k = -\varrho s\,\nabla T + \nabla p. \tag{3.29}$$

Other thermostatic relations will not be given here, but it should be emphasized that the hypothesis of cellular equilibrium is equivalent to the condition that all relations of equilibrium thermodynamics are considered valid for the corresponding infinitesimal mass elements of a non-equilibrium system.

The postulate of local equilibrium evidently represents a limitation to the validity range of the theory developed. From the macroscopic viewpoint, the domain of validity of the theory can only be estimated by comparison of the conclusions drawn from the theory with experience. Earlier it was thought that the hypothesis of local equilibrium is a very strong requirement and is justified only in case of systems which are very close to the equilibrium state. Nowadays, due to the successes of irreversible thermodynamics on the one hand, and on the basis of Meixner's [33], Prigogine's [34] and Reik's [35] kinetic examinations on the other hand, we rely upon the reality of the local equilibrium also in case of systems more distant from equilibrium. Without giving detailed description of these examinations, let us cite Meixner's results. Meixner has stated that the volume elements of a mono-atomic gas system can be considered as equilibrium cells as far as any variations in temperature and velocity within the interval of the mean free path are small with respect to the average absolute temperature and sound velocity. For a gas of normal state this means that it can be considered as a system in cellular equilibrium up to a temperature gradient of about 10^5 C° cm^{-1}, i.e. the methods of irreversible thermodynamics can be employed up to this limit. Today we think that, with the exception of turbulent phenomena, shock waves and rapid plasma processes, the hypothesis of the cellular equilibrium may be accepted. Of course, from the viewpoint of field theory, the applicability of the condition of cellular equilibrium for singularities (curves, surfaces) can depend a priori on many other factors, by which the applicability limits of the theory may be influenced. Here we allude for example to the boundary-layer of semiconductors etc. Nevertheless Meixner's example and the results of irreversible thermodynamics prove that the hypothesis of the local equilibrium—disregarding extreme cases—is a realistic assumption.

2. Conservation of Energy and the Balances of Internal Energy

Our task is to determine the actual forms of the internal energy balances (3.6) and (3.7). Since the balances of mechanical energy were derived in the previous chapter [(2.189) and (2.194)] and the global form of the conservation of total energy in term of internal energy was formulated in (3.4), it can be concluded that the balances of the internal energy can be derived from the conservation of the total energy with the aid of the balances of mechanical energy. In such cases, however, the classification of the diffusion kinetic energy (2.173) gives rise to particular difficulties, because the definition of the total specific energy (at least at first sight) can be given in two equivalent forms.

Let us define the total specific energy in the usual way as the sum of the specific mechanical energy (2.188) and the specific internal energy. Hence, let us consider the quantity

$$\varepsilon \equiv \varepsilon_{\rm m} + u = \frac{1}{2}\boldsymbol{v}^2 + \frac{1}{2}\Theta\boldsymbol{\omega}^2 + \varphi + u \tag{3.30}$$

as a total specific energy, which satisfies the

$$\frac{\partial \varrho\varepsilon}{\partial t} + \nabla\cdot\boldsymbol{J}^0_\varepsilon = 0 \tag{3.31}$$

sourceless local balance expressing the conservation of energy [3, 4, 8, 13, 17, 26, 27]. Substracting the corresponding sides of (2.194) from (3.31) the balance

$$\frac{\partial \varrho(\varepsilon - \varepsilon_{\rm m})}{\partial t} + \nabla\cdot(\boldsymbol{J}^0_\varepsilon - \boldsymbol{J}^0_{\varepsilon_{\rm m}}) = -\sigma_{\varepsilon_{\rm m}} \tag{3.32}$$

is obtained. By comparing it to (3.6), (3.30) and (2.195) it will be seen that (3.32) is a local internal energy balance with the

$$\boldsymbol{J}^0_{\rm u} = \boldsymbol{J}^0_\varepsilon - \boldsymbol{J}^0_{\varepsilon_{\rm m}} = \boldsymbol{J}^0_\varepsilon - \sum_{k=1}^{K}\varphi_k\boldsymbol{J}_k - \mathbf{P}\cdot\boldsymbol{v} - \varrho\varepsilon_{\rm m}\boldsymbol{v} \tag{3.33}$$

local internal energy current density and with the source density

$$\sigma_{\rm u} = -\sigma_{\varepsilon_{\rm m}} = -\widetilde{\mathbf{P}}:\nabla\boldsymbol{v} + 2\boldsymbol{\omega}\cdot\boldsymbol{P}^{\rm va} + \sum_{k=1}^{K}\boldsymbol{J}_k\cdot\boldsymbol{F}_k - \sum_{k=1}^{K}\sum_{j=1}^{R}\varphi_k\nu_{kj}J_j. \tag{3.34}$$

If it is assumed that during chemical reactions the condition (2.185) fulfilled, and if the part of $\boldsymbol{J}^0_{\rm u}$ separated from the convective current density $\varrho u\boldsymbol{v}$ of the internal energy is called *heat current*:

$$\boldsymbol{J}_{\rm q} = \boldsymbol{J}^0_{\rm u} - \varrho u\boldsymbol{v} = \boldsymbol{J}^0_\varepsilon - \sum_{k=1}^{K}\varphi_k\boldsymbol{J}_k - \mathbf{P}\cdot\boldsymbol{v} - \varrho\varepsilon\boldsymbol{v} \tag{3.35}$$

then for the internal energy the local balance

$$\frac{\partial \varrho u}{\partial t} + \nabla\cdot(\boldsymbol{J}_{\rm q} + \varrho u\boldsymbol{v}) = -\widetilde{\mathbf{P}}:\nabla\boldsymbol{v} + 2\boldsymbol{\omega}\cdot\boldsymbol{P}^{\rm va} + \sum_{k=1}^{K}\boldsymbol{J}_k\cdot\boldsymbol{F}_k \tag{3.36}$$

is obtained from (3.32) with (2.191).

This balance equation represents the first law of thermodynamics in local form. We approach, however, the more usual form of the first law by transforming (3.36) with (2.21) to the substantial form. In the above-mentioned manner, with the choice $a \equiv u$ in (2.21), we get from (3.36) that

$$\varrho\dot{u} + \nabla\cdot\boldsymbol{J}_{\rm q} = -\widetilde{\mathbf{P}}:\nabla\boldsymbol{v} + 2\boldsymbol{\omega}\cdot\boldsymbol{P}^{\rm va} + \sum_{k=1}^{K}\boldsymbol{J}_k\cdot\boldsymbol{F}_k. \tag{3.37}$$

The comparison of this substantial balance to (2.15) shows that the heat flow introduced in (3.35) is identical with the substantial current density of the internal energy:

$$\boldsymbol{J}_{\mathrm{q}} \equiv \boldsymbol{J}_{\mathrm{u}}. \tag{3.38}$$

This relation must be considered to be the exact definition of the concept of heat flow, fairly loosely used in experimental physics and heat technics.

The balance equation (3.37) of the internal energy can be easily converted after a small simplification and transformation in the source term σ_{u} in a somewhat more transparent form of the first law. Indeed, confining ourselves to the case of a symmetric pressure tensor, and relying only upon the splitting (2.94), we get

$$\sigma_{\mathrm{u}} = -\mathbf{P}:\nabla \boldsymbol{v} + \sum_{k=1}^{K} \boldsymbol{J}_k \cdot \boldsymbol{F}_k = -p\nabla \cdot \boldsymbol{v} - \mathbf{P}^{\mathrm{v}}:\nabla \boldsymbol{v} + \sum_{k=1}^{K} \boldsymbol{J}_k \cdot \boldsymbol{F}_k \tag{3.39}$$

which includes, of course, already less than (3.34). Writing now the conservation of mass (2.17) with the specific volume $v = \varrho^{-1}$ instead of ϱ, we have

$$\varrho \dot{v} - \nabla \cdot \boldsymbol{v} = 0, \tag{3.40}$$

and requiring that the heat current $\boldsymbol{J}_{\mathrm{q}}$, according to the idea of the "heat fluid theory", shall satisfy the balance equation

$$\varrho \dot{q} + \nabla \cdot \boldsymbol{J}_{\mathrm{q}} = 0 \tag{3.41}$$

expressing the conservation of "heat fluid" (where $\mathrm{d}q = \dot{q}\,\mathrm{d}t$ is the heat supplied per unit mass), the internal energy balance (3.37) can be given with (3.39), (3.40) and (3.41) in the form

$$\dot{u} = \dot{q} - p\dot{v} - v\,\mathbf{P}^{v}:\nabla \boldsymbol{v} + v \sum_{k=1}^{K} \boldsymbol{J}_k \cdot \boldsymbol{F}_k. \tag{3.42}$$

Concerning the foregoings it must be kept in mind that the identity (3.38) represents the exact definition of the heat flow $\boldsymbol{J}_{\mathrm{q}}$. The sourceless heat balance (3.41) serves only to give the balance equation (3.36) and (3.37) of the internal energy in such a form as (3.42) which corresponds better to the concepts used in connection with the formulation of the first law in experimental physics. Indeed, by comparing equation (3.8) expressing the first law to the internal energy balance (3.42), it can be seen that the latter is nothing else than the first law in local form, by which the variation of the specific internal energy is determined as the sum of the supplied heat $\mathrm{d}q$ and of the performed work

$$\mathrm{d}w = \left(-p\dot{v} - v\,\mathbf{P}^{\mathrm{v}}:\nabla \boldsymbol{v} + v \sum_{k=1}^{K} \boldsymbol{J}_k \cdot \boldsymbol{F}_k\right) \mathrm{d}t. \tag{3.43}$$

Here the first term on the right-hand side is identical with the volume work per unit mass of the continuum. The second term describes the work performed against friction. Finally, the sum given in the last term defines the work of the external forces. Owing to the fact that this term can also be given by using (1.42) and (2.84) in the form

$$v\left(\sum_{k=1}^{K} \boldsymbol{J}_k \cdot \boldsymbol{F}_k\right) \mathrm{d}t = v\left(\sum_{k=1}^{K} \varrho_k \boldsymbol{w}_k \cdot \boldsymbol{F}_k\right) \mathrm{d}t = v\left(\sum_{k=1}^{K} \varrho_k \boldsymbol{F}_k \cdot \boldsymbol{v}_k - \varrho \boldsymbol{F} \cdot \boldsymbol{v}\right) \mathrm{d}t \tag{3.44}$$

it can be seen that the sum given in the last therm of (3.43) determines the work of diffusion with respect to the barycenter. The expression (3.44) shows that this diffusion work is the difference of the total work expressed by the first sum and of the work of the resultant external force $\boldsymbol{F}$ acting upon the barycenter. This must be so, because the work $(\varrho \boldsymbol{F} \cdot \boldsymbol{v})\, \mathrm{d}t$ acting upon the barycenter cannot be counted to the variation of the internal energy of the continuum, because, according to relation (2.147), the latter is determined by the external source density of the barycentric translational kinetic energy.

Let us return to the problem raised earlier concerning the total energy of multi-component continua. First of all, let us emphasize, that in the case of multi-component continua the energy quantity ε defined in (3.30) is not equal to the total specific energy of such continua, because ε and ε_m do not contain the kinetic energy of diffusion ε_d defined in (2.173). Hence, the quantity ε cannot be considered to be the total specific energy for the same reason for which the energy ε_m is not identical to the total specific mechanical energy with the exception of the particular case of single-component continua. From the foregoings it is evident that when referring to multi-component continua, the quantity

$$\varepsilon^\mathrm{t} \equiv \varepsilon_\mathrm{m}^\mathrm{t} + u = \frac{1}{2} \boldsymbol{v}^2 + \frac{1}{2} \Theta \boldsymbol{\omega}^2 + \frac{1}{2} \sum_{k=1}^{K} c_k \boldsymbol{w}_k^2 + \varphi + u \tag{3.45}$$

must be considered as total specific energy instead of ε because the kinetic energy of diffusion is also incorporated.

It can be seen that the two possibilities of choosing of the total specific energy given in (3.30) and (3.45), respectively, is due to the two different choices (2.188) and (2.196) of the total specific mechanical energy. Evidently, the quantities ε_m and ε, in which the kinetic energy of diffusion is not incorporated, cannot be considered as total specific energy quantities in the case of multi-component continua. It is not a matter of free choice, but rather an important problem that ε or ε^t is the total specific energy quantity for which the existence of the sourceless balance equation expressing the conservation of energy is required.

Therefore all derivations which have been given in the literature [3, 4, 8, 13, 17, 26, 27] and which are based upon the balance equation expressing the conservation (3.31), required for the specific energy ε, must be considered incomplete, if the following are taken into account, according to which the majority of the examinations based upon (3.31) remains in essence unchanged even in the case of a treatment which is correct in principle. Nevertheless, in the case of multi-component systems, the treatment given up to now does not give complete information on the energy relations and as it can be misleading it should be completed and corrected. The necessary completion and correction will be carried out in the following.

It is evident that the adequate internal energy balances can in principle be correctly obtained, if the existence of the sourceless balance equation

$$\frac{\partial \varrho \varepsilon^{t}}{\partial t} + \nabla \cdot \boldsymbol{J}^{0}_{\varepsilon_t} = 0 \tag{3.46}$$

expressing the conservation of energy is required for the total energy ε^{t} defined in (3.45). From this we get the local balance of the internal energy by using (3.45) and (2.198):

$$\frac{\partial \varrho(\varepsilon^{t} - \varepsilon^{t}_{\mathrm{m}})}{\partial t} + \nabla \cdot (\boldsymbol{J}^{0}_{\varepsilon_t} - \boldsymbol{J}^{0}_{\varepsilon^{t}_{\mathrm{m}}}) = -\sigma_{\varepsilon_{\mathrm{m}}} = \sigma_{\mathrm{u}}. \tag{3.47}$$

This deviates from (3.32) only inasmuch as the specific internal energy u is now defined as

$$u = \varepsilon^{t} - \varepsilon^{t}_{\mathrm{m}} \tag{3.48}$$

following from (3.45), whereas the corresponding local current density is described by

$$\boldsymbol{J}^{0}_{\mathrm{u}} = \boldsymbol{J}^{0}_{\varepsilon^{t}} - \boldsymbol{J}^{0}_{\varepsilon^{t}_{\mathrm{m}}}. \tag{3.49}$$

From (2.198) and (3.49) we have

$$\boldsymbol{J}^{0}_{\varepsilon^{t}} = \sum_{k=1}^{K} \varphi_k \boldsymbol{J}_k + \mathbf{P} \cdot \boldsymbol{v} + \varrho \varepsilon^{t}_{\mathrm{m}} \boldsymbol{v} + \boldsymbol{J}^{0}_{\mathrm{u}}, \tag{3.50}$$

for the local current density of the total energy, which still contains, owing to the definition (2.196) of $\varepsilon^{t}_{\mathrm{m}}$, the convective current density $\varrho \varepsilon_{\mathrm{d}} \boldsymbol{v}$ of the kinetic energy of diffusion (2.173) with respect to motion of the centre of gravity.

Practically the relations (3.45) to (3.50) are equivalent to the similar relations (3.30) to (3.35). However, it is of theoretical importance that the conservation of energy must be given for the real total specific energy ε^{t} instead of ε. It is remarkable that the definition (3.30) of the specific internal energy is in agreement with the definition (3.45) (or

(3.48)) as a result of forming of difference. Any definition given for the internal energy interpret such a specific quantity of energy, which does not contain the kinetic energy of diffusion corresponding to the macroscopic diffusional motion of the components. In other words, the specific internal energy introduced by us depends in both cases on the microparameters only and includes contributions from the thermal agitation and the short-range corpuscular interactions in accordance with the usual interpretation of the internal energy.

It has been mentioned already that generally in literature when determining the balance equation of internal energy, the conservation law (3.31) required for the energy ε is taken as a basis [3, 4, 8, 13, 17, 26, 27]. The various authors were evidently biased by the fact that the mechanical balance equations determined earlier (for example (2.189) or (2.194) refer to the quantity ε_m, and thus a balance equation for the internal energy can be obtained only with the use of them. In other words, until now we were obliged to confine ourselves to the mechanical balances determined for the energy quantities containing only the barycentric kinetic energy $\varepsilon_t \equiv \frac{1}{2}\boldsymbol{v}^2$ and not including the kinetic energy of diffusion. Indeed, for the total specific translational kinetic energy $\overset{t}{\varepsilon}_t$ defined in (2.170) the exact balance equation cannot be given in a direct way, only in the case derived from the impulse balances of the type (2.76). However, such a direct derivation is not known yet. Therefore, though the relations given in expressions (3.45) to (3.50) and completing the usual ones yield a more complete (and in principle correct) picture on the balance of the internal energy than the hitherto employed ones, further research is required in this respect. The fact that the uncleared state of the conditions may be misleading shall be presented in the following instances.

Let us consider the quantity ε defined in (3.30) as a total energy. Then, as the difference of (3.30) and the specific mechanical energy of (2.196) we arrive at a definition of the specific internal energy:

$$\varepsilon - \overset{t}{\varepsilon}_m = -\frac{1}{2}\sum_{k=1}^{K} c_k \boldsymbol{w}_k^2 + u = u^*. \tag{3.51}$$

Similarly, (3.31) and (2.188) lead again to another definition of the internal energy:

$$\varepsilon^t - \varepsilon_m = \frac{1}{2}\sum_{k=1}^{K} c_k \boldsymbol{w}_k^2 + u = u^{**}. \tag{3.52}$$

Though it is certain that no correct internal energies are represented by the quantities u^* and u^{**} it is shown by the foregoing that more care must be taken while giving the local interpretation of internal energy. Indeed, in the otherwise excellent monographs of de Groot and Mazur [3] we find that these authors give two different definitions for the specific internal energy: first, that following from (3.30)

$$\varepsilon - \varepsilon_m = u \tag{3.53}$$

which is correct and equivalent to (3.48) (or rather to (3.30)), and secondly, (3.51) which is incorrect. In fact the above-mentioned authors overlooked, that if the kinetic energy of diffusion is included in the total mechanical energy, —as it should be done—the kinetic energy of diffusion should also be taken into account in the total energy. This we have done in the definition of (3.45). Though we are of the opinion that on the basis of the ideas explained above we are nearer to the description of the correct internal energy balances of multicomponent systems, it must be emphasized that further investigations are required in this field.

It is shown by the balances (3.36), (3.37) and also by (3.42), expressing the first law in the local form, that in general the internal energy is not conserved. If, for example, the source density σ_u of the substantial balance (3.37) is written in detail on the basis of the relation $\sigma_u = -\sigma_{\varepsilon_m}$ following from (3.32), and further taking (2.191) and (2.185) into account, the source density

$$\sigma_u = -(p + p^v)\,\nabla \cdot \boldsymbol{v} - \overset{0}{\mathbf{P}}{}^{vs} : (\overset{0}{\nabla \boldsymbol{v}})^s - \boldsymbol{P}^{va} \cdot (\nabla \times \boldsymbol{v} - 2\boldsymbol{\omega}) + \sum_{k=1}^{K} \boldsymbol{J}_k \cdot \boldsymbol{F}_k \tag{3.54}$$

is obtained. In this expression

$$\sigma_u^i = -(p + p^v)\,\nabla \cdot \boldsymbol{v} - \overset{0}{\mathbf{P}}{}^{vs} : (\overset{0}{\nabla \boldsymbol{v}})^s - \boldsymbol{P}^{va} \cdot (\nabla \times \boldsymbol{v} - 2\boldsymbol{\omega}) \tag{3.55}$$

is the "internal", whereas

$$\sigma_u^e = \sum_{k=1}^{K} \boldsymbol{J}_k \cdot \boldsymbol{F}_k \tag{3.56}$$

is the "external" source density in accordance with (2.192) and (2.193), since evidently $\sigma_u^i = -\sigma_{\varepsilon_m}^i$ and $\sigma_u^e = -\sigma_{\varepsilon_m}^e$.

It is worthwhile to note that in the "internal" source density (3.55), with the exception of the term $p\nabla \cdot \boldsymbol{v} = \varrho p \dot{v}$ related to the volume work, all other terms are related to the work of the viscous forces. Thus, it can be seen from (3.55) that in the particular case of Joule's classical experiment with the rotating paddles, how the macroscopic kinetic energy of the masses turning the paddles is transformed through the work of the viscous forces into the submicroscopic kinetic energy of the molecules of the liquid, i.e. into the internal energy.

3. Entropy Balances and Entropy Production

In possession of the given balance equations we have a possibility for the determination of the entropy balances playing a central rôle in non-equilibrium thermodynamics. Let us determine an actual form of the entropy balances (3.16) for the case of a very general model of multi-component hydro-thermodynamic systems. Substituting $\dot{u}$ from (3.37) into the Gibbs relation (3.19) and similarly eliminating $\dot{c}_k$ from it with the component balances (2.46), we get

$$\varrho \dot{s} + \frac{\nabla \cdot \boldsymbol{J}_{\mathrm{q}}}{T} - \sum_{k=1}^{K} \frac{\mu_k}{T} \nabla \cdot \boldsymbol{J}_k$$

$$= \frac{-\sum_{j=1}^{R} \sum_{k=1}^{K} \mu_k \nu_{kj} J_j + \sum_{k=1}^{K} \boldsymbol{J}_k \cdot \boldsymbol{F}_k - \widetilde{\mathbf{P}} : \nabla \boldsymbol{v} + 2\boldsymbol{\omega} \cdot \boldsymbol{P}^{\mathrm{va}}}{T}. \tag{3.57}$$

This equation does not correspond to the substantial entropy balance of the type (3.16). If, however, the divergences are transformed by

$$\frac{1}{T} \nabla \cdot \boldsymbol{J}_{\mathrm{q}} = \nabla \cdot \frac{\boldsymbol{J}_{\mathrm{q}}}{T} - \boldsymbol{J}_{\mathrm{q}} \cdot \nabla \frac{1}{T} \tag{3.58}$$

and

$$\frac{\mu_k}{T} \nabla \cdot \boldsymbol{J}_k = \nabla \cdot \frac{\mu_k \boldsymbol{J}_k}{T} - \boldsymbol{J}_k \cdot \nabla \frac{\mu_k}{T}, \qquad (k = 1, 2, \ldots, K) \tag{3.59}$$

further, by giving the last two terms of (3.57) in detail according to (3.54) for the source density σ_{u}, the substantial entropy balance

$$\varrho \dot{s} + \nabla \cdot \frac{\boldsymbol{J}_{\mathrm{q}} - \sum_{k=1}^{K} \mu_k \boldsymbol{J}_k}{T} \tag{3.60}$$

$$= \frac{\sum_{j=1}^{R} J_j A_j^* + \boldsymbol{J}_{\mathrm{q}} \cdot \boldsymbol{X}_{\mathrm{q}}^* + \sum_{k=1}^{K} \boldsymbol{J}_k \cdot \boldsymbol{X}_k^* + p^{\mathrm{v}} X_{\mathrm{v}}^* + \overset{0}{\mathbf{P}}{}^{\mathrm{vs}} : \overset{0}{\mathbf{X}}{}_{\mathrm{v}}^{\mathrm{s}*} + \boldsymbol{P}^{\mathrm{va}} \cdot \boldsymbol{X}_{\mathrm{v}}^{\mathrm{a}*}}{T}$$

is obtained. Here the so-called thermodynamic forces have been introduced by the quantities A_j^*, $\boldsymbol{X}_{\mathrm{q}}^*$, $\boldsymbol{X}_k^*$, X_{v}^*, $\overset{0}{\mathbf{X}}{}_{\mathrm{v}}^{\mathrm{s}*}$, $\boldsymbol{X}_{\mathrm{v}}^{\mathrm{a}*}$. By definition, the interpretation of these is, the following:

The chemical affinity of the j-th reaction,

$$A_j^* \equiv -\sum_{k=1}^{K} \mu_k \nu_{kj}, \qquad (j = 1, 2, \ldots, R) \tag{3.61}$$

is a scalar thermodynamic force conjugated to the scalar J_j, which is the rate of the j-th chemical reaction.

The polar vector

$$\boldsymbol{X}_{\mathrm{q}}^* \equiv -\frac{\nabla T}{T} = -\nabla \ln T \tag{3.62}$$

is a thermodynamic force causing the heat conductions phenomena and conjugated to the polar vector of heat current density $\boldsymbol{J}_{\mathrm{q}}$. This force is sometimes denoted by $\boldsymbol{X}_{\mathrm{u}}^*$ as well [4], owing to the general definition of the heat current density (3.38).

The thermodynamic forces of diffusion, also containing the arbitrary external forces $\boldsymbol{F}_k$, are the quantities

$$\boldsymbol{X}_k^* \equiv \boldsymbol{F}_k - T\,\nabla\left(\frac{\mu_k}{T}\right), \qquad (k = 1, 2, \ldots, K) \tag{3.63}$$

conjugated to the diffusion current density $\boldsymbol{J}_k$.

The scalar viscous force

$$X_{\mathrm{v}}^* \equiv -\nabla\cdot\boldsymbol{v} \tag{3.64}$$

is conjugated to the viscous pressure p^{v} as to a scalar impulse flow, by which the phenomena related to the volume viscosity of the compressible fluids are caused.

The tensorial viscous force

$$\overset{0}{\mathbf{X}}{}_{\mathrm{v}}^{\mathrm{s}*} \equiv -\,(\overset{0}{\nabla\boldsymbol{v}})^{\mathrm{s}} \tag{3.65}$$

is conjugated to the symmetric part of the second order viscous pressure tensor of zero trace $\overset{0}{\mathbf{P}}{}^{\mathrm{vs}}$ and which gives rise to the phenomena of shear viscosity.

Finally, the axial vector

$$\boldsymbol{X}_{\mathrm{v}}^{\mathrm{a}*} \equiv -\,(\nabla\times\boldsymbol{v} - 2\boldsymbol{\omega}) \tag{3.66}$$

is conjugated to the axial vector $\boldsymbol{P}^{\mathrm{va}}$ formed by the antisymmetric part of the viscous pressure tensor, by which also a thermodynamic force causing an irreversible process is represented, at least in every fluid system in which the conservation law of the angular momentum is not fulfilled.

Comparing (3.60) to the general balance (3.16) it can be seen that the substantial entropy current density is

$$\boldsymbol{J}_{\mathrm{s}} = \frac{\boldsymbol{J}_{\mathrm{q}} - \sum_{k=1}^{K} \mu_k \boldsymbol{J}_k}{T}, \tag{3.67}$$

whereas the entropy production σ per unit volume and unit time or the so-called energy dissipation is determined by

$$T\sigma = \sum_{j=1}^{R} J_j A_j^* + \boldsymbol{J}_{\mathrm{q}}\cdot\boldsymbol{X}_{\mathrm{q}}^* + \sum_{k=1}^{K} \boldsymbol{J}_k\cdot\boldsymbol{X}_k^* + p^{\mathrm{v}} X_{\mathrm{v}}^* + \overset{0}{\mathbf{P}}{}^{\mathrm{vs}} : \overset{0}{\mathbf{X}}{}_{\mathrm{v}}^{\mathrm{s}*} + \boldsymbol{P}^{\mathrm{va}}\cdot\boldsymbol{X}_{\mathrm{v}}^{\mathrm{a}*}. \tag{3.68}$$

This energy quantity is dissipated in a multi-component, reacting, non-isothermal and viscous hydro-thermodynamic system per unit volume and unit time, if the system is a seat of the irreversible processes in question. On the other hand, this is the energy quantity which has already been recognized in more simple particular cases and was called *non-compensated heat* by CLAUSIUS [5]. According to inequalities (3.11) and (3.14b) expressing the second law of irreversible transformations, the energy dissipation $T\sigma$, and also the entropy production

$$\sigma = \sum_{j=1}^{R} J_j \frac{A_j^*}{T} + \boldsymbol{J}_q \cdot \frac{\boldsymbol{X}_q^*}{T} + \sum_{k=1}^{K} \boldsymbol{J}_k \cdot \frac{\boldsymbol{X}_k^*}{T} + p^v \frac{X_v^*}{T} + \overset{0}{\mathbf{P}}{}^{vs} : \frac{\overset{0}{\mathbf{X}}{}_v^{s*}}{T} + \boldsymbol{P}^{va} \cdot \frac{\boldsymbol{X}_v^{a*}}{T} \geqq 0 \tag{3.69}$$

must be positive definite quantities.

Now let us mention an important representation problem. The energy dissipation $T\sigma$ is a positive definite quantity which is a local measure of irreversibility and which is determined by an adequate number of thermodynamic currents J_j, $\boldsymbol{J}_q$, $\boldsymbol{J}_k$, p^v, $\overset{0}{\mathbf{P}}{}^{vs}$, $\boldsymbol{P}^{va}$ and conjugated forces (3.61) to (3.66), according to the bilinear form (3.68). The thermodynamic forces (3.61) to (3.66) determine directly the dissipation of energy $T\sigma$, and thus, if we intend to determine directly the expression of the entropy production, it is advisable to introduce new quantities

$$A_j \equiv \frac{A_j^*}{T} = -\sum_{k=1}^{K} \frac{\mu_k}{T} \nu_{kj}, \qquad (j = 1, 2, \ldots, R) \tag{3.70}$$

$$\boldsymbol{X}_q \equiv \frac{\boldsymbol{X}_q^*}{T} = -\frac{1}{T^2} \nabla T = \nabla\left(\frac{1}{T}\right), \tag{3.71}$$

$$\boldsymbol{X}_k \equiv \frac{\boldsymbol{X}_k^*}{T} = \frac{\boldsymbol{F}_k}{T} - \nabla\left(\frac{\mu_k}{T}\right), \qquad (k = 1, 2, \ldots, K) \tag{3.72}$$

$$X_v \equiv \frac{X_v^*}{T} = -\frac{1}{T} \nabla \cdot \boldsymbol{v}, \tag{3.73}$$

$$\overset{0}{\mathbf{X}}{}_v^s \equiv \frac{\overset{0}{\mathbf{X}}{}_v^{s*}}{T} = -\frac{1}{T} (\overset{0}{\nabla \boldsymbol{v}})^s, \tag{3.74}$$

$$\boldsymbol{X}_v^a \equiv \frac{\boldsymbol{X}_v^{a*}}{T} = -\frac{1}{T} (\nabla \times \boldsymbol{v} - 2\boldsymbol{\omega}) \tag{3.75}$$

as thermodynamic forces by which the entropy production can be given as

$$\sigma = \sum_{j=1}^{R} J_j A_j + \boldsymbol{J}_q \cdot \boldsymbol{X}_q + \sum_{k=1}^{K} \boldsymbol{J}_k \cdot \boldsymbol{X}_k + p^v X_v + \overset{0}{\mathbf{P}}{}^{vs} : \overset{0}{\mathbf{X}}{}_v^s + \boldsymbol{P}^{va} \cdot \boldsymbol{X}_v^a \geqq 0. \tag{3.76}$$

In the following, the forces marked with * will be called thermodynamic forces belonging to the energy dissipation $T\sigma$. Thus, if in the course of our calculations we have to work with them and directlv determine the energy dissipation, it can be said that we are working in *energy picture*. Whereas, when we intend to determine directly σ by using the force (3.70) to (3.75), we said that we work in *entropy picture.*[1] Both pictures have their advantages and drawbacks in the case of the given problems, and evidently these two representations can be transformed into one another with the application of the formulae (3.70) to (3.75). Therefore, until we are not compelled by some particular reason, we confine ourselves to the analysis of one of the representations. In the following we prefer the entropy picture (3.76).

According to (3.76) the entropy production includes four source terms corresponding to irreversible phenomena of considerably different physical nature, which are determined by bilinear forms of currents and forces of different tensorial orders. Thus the entropy source due to chemical reactions

$$\sigma_{\mathrm{c}} \equiv \sum_{j=1}^{R} J_j A_j \qquad (3.77)$$

is determined by the sum of bilinear forms of the scalars J_j, and A_j. The entropy production due to heat conduction is determined by the two factors $\boldsymbol{J}_{\mathrm{q}}$ and $\boldsymbol{X}_{\mathrm{q}}$ as polar vectors:

$$\sigma_{\mathrm{q}} \equiv \boldsymbol{J}_{\mathrm{q}} \cdot \boldsymbol{X}_{\mathrm{q}}. \qquad (3.78)$$

Similarly, the entropy production due to diffusion is the sum of the scalar products of the polar vectors $\boldsymbol{J}_k$ and $\boldsymbol{X}_k$, i.e.

$$\sigma_{\mathrm{d}} \equiv \sum_{k=1}^{K} \boldsymbol{J}_k \cdot \boldsymbol{X}_k. \qquad (3.79)$$

Finally, the entropy production due to viscous phenomena

$$\sigma_{\mathrm{v}} \equiv p^{\mathrm{v}} X_{\mathrm{v}} + \overset{0}{\mathbf{P}}{}^{\mathrm{vs}} : \overset{0}{\mathbf{X}}{}^{\mathrm{s}}_{\mathrm{v}} + \boldsymbol{P}^{\mathrm{va}} \cdot \boldsymbol{X}^{\mathrm{a}}_{\mathrm{v}} \qquad (3.80)$$

is determined as a bilinear form of the impulse current densities and corresponding thermodynamic forces related to viscous phenomena of volume, shear and internal rotation. The total entropy production (3.76) valid in the case of the simultaneous process of the above four phenomena can be considered as the sum of the partial entropy productions. Consequently, σ can be considered, in general, as the bilinear expression of f independent scalar fluxes J_i and scalar forces X_i, i.e.

$$\sigma = \sum_{i=1}^{f} J_i X_i. \qquad (3.81)$$

---

[1] The usefulness of these new denominations for the different representations will particularly be seen in Chapters V and VI.

On first sight it seems that in the treated partial entropy productions of (3.81) a number $f = R + 3 + 3K + 1 + 5 + 3$ of independent scalar flux and force components must occur. However, this is not the case, because the relation (1.43) always exists as a local constraint for the diffusion current densities $\boldsymbol{J}_k$, by the aid of which any component, for example the current density $\boldsymbol{J}_K$ of the K-th component, can be eliminated from (3.79). Hence, expressing the partial entropy production due to diffusion by independent current densities and forces, we have

$$\sigma_{\mathrm{d}} \equiv \sum_{k=1}^{K-1} \boldsymbol{J}_k \cdot \boldsymbol{X}'_k \tag{3.82}$$

where

$$\boldsymbol{X}'_k \equiv \frac{\boldsymbol{F}_k - \boldsymbol{F}_K}{T} - \nabla\left(\frac{\mu_k - \mu_K}{T}\right), \qquad (k = 1, 2, \ldots, K-1) \tag{3.83}$$

are the diffusion forces conjugated to the independent current densities. It is shown by (3.82) that the number of the bilinear sums formed from the Cartesian components of the independent current densities and forces in (3.81) is reduced to $f = R + 3 + 3(K - 1) + 1 + 5 + 3$ owing to condition (1.43).

4. The Linear Kinematical Constitutive Equations

In thermodynamic equilibrium, the entropy production σ must be vanish, so that all the independent scalar force components and also the scalar flux components conjugated to them become simultaneously zero. This condition, and at the same time, the most general interactions between the independent fluxes and forces are expressed—in linear order—by Onsager's linear kinematical constitutive equations (or laws):

$$J_i = \sum_{k=1}^{f} L_{ik} X_k, \qquad (i = 1, 2, \ldots, f). \tag{3.84}$$

Here the scalars J_i and X_k denote the independent scalar thermodynamic fluxes and forces, whereas in the case of vector and tensor processes they mean all the Cartesian components of the correspondent quantities of tensorial order which occur in the bilinear expression (3.76) of the entropy production.

In the following the scalar flows (rates) and in case of vector and tensor current densities the scalar components of them will often be called fluxes. Since in (3.76), valid for multi-component and reacting hydro-thermodynamic systems, the number of the scalar components of the independent currents and forces is $f = R + 3 + 3(K - 1) + 1 + 5 + 3$, and consequently the square matrix of the conductivity coefficients contains $f^2 = R^2 + 9K^2 + 6KR + 18R + 54K + 81$ scalar elements. Of course, the L_{ik} Onsager coefficients are functions of the local state parameters, such as temperature, pressure, chemical potentials of concentrations, or possibly magnetic field strength, etc. However, coefficients are considered constant in the linear theory with respect to the fluxes and forces occuring in the linear constitutive equations (3.84), i.e. with respect to the gradients of the local state parameters.

It is maintained that the foregoings are also valid for the fluxes and forces of (3.68) in which case the linear constitutive equations

$$J_i = \sum_{k=1}^{f} L_{ik}^* X_k^*, \qquad (i = 1, 2, \ldots, f) \tag{3.85}$$

belonging to the energy picture must be used. Between Onsager's coefficients of the linear relations (3.84) given in the entropy picture, and (3.85), given for the energy representation owing to (3.70) to (3.75), referring to the forces, the relations

$$L_{ik} = TL_{ik}^*, \qquad (i, k = 1, 2, \ldots, f) \tag{3.86}$$

must be valid, because the fluxes are identical in both pictures. Now, the actual forms of the linear kinematical constitutive equations (3.84) will be given for systems displaying a quite arbitrary anisotropy and in the following also for the case of isotropic bodies.

a) Anisotropic Case. The majority of the hydro-thermodynamic systems is isotropic; however, sometimes for the sake of generality, but also with respect to particular systems (for example, ionized gas plasma in magnetic fields), it may be important to give the actual form of the linear kinematical constitutive equations for general anisotropic systems. Let us give the final form of the entropy production by making use of relations (3.82) and (3.83):

$$\sigma = \sum_{j=1}^{R} J_j A_j + p^{\mathrm{v}} X_{\mathrm{v}} + \boldsymbol{J}_{\mathrm{q}} \cdot \boldsymbol{X}_{\mathrm{q}} + \sum_{k=1}^{K-1} \boldsymbol{J}_k \cdot \boldsymbol{X}_k' + \boldsymbol{P}^{\mathrm{va}} \cdot \boldsymbol{X}_{\mathrm{v}}^{\mathrm{a}} + \overset{0}{\mathbf{P}}{}^{\mathrm{vs}} : \overset{0}{\mathbf{X}}{}_{\mathrm{v}}^{\mathrm{s}} \geqq 0 \tag{3.87}$$

which contains only independent parameters. In this expression the partial entropy productions related to the individual irreversible processes were arranged in such a manner that the tensorial order of the currents and forces determining them are increasing from left to right. By taking the tensorial order of the fluxes and forces as well as their polar and axial character into account, and taking into consideration the interaction of all the Cartesian components of the independent currents and forces in the description of the cross-effects represented by the coefficients of mixed indices, the actual scheme of the linear constitutive equations (3.84) is the following:

$$J_j = \sum_{r=1}^{R} \overset{(ss)}{L_{jr}^{\mathrm{cc}}} A_r + \overset{(ss)}{L_j^{\mathrm{cv}}} X_{\mathrm{v}} + \overset{(sv)}{\boldsymbol{L}_j^{\mathrm{cq}}} \cdot \boldsymbol{X}_{\mathrm{q}} + \sum_{k=1}^{K-1} \overset{(sv)}{\boldsymbol{L}_{jk}^{\mathrm{cd}}} \cdot \boldsymbol{X}_k' + \overset{(s\boldsymbol{a})}{\boldsymbol{L}_j^{\mathrm{cv}}} \cdot \boldsymbol{X}_{\mathrm{v}}^{\mathrm{a}} + \overset{(s\mathbf{t})}{\mathbf{L}_j^{\mathrm{cv}}} : \overset{0}{\mathbf{X}_{\mathrm{v}}^{\mathrm{s}}},$$

$$(j = 1, 2, \ldots, R)$$

$$p^{\mathrm{v}} = \sum_{r=1}^{R} \overset{(ss)}{L_r^{\mathrm{vc}}} A_r + \overset{(ss)}{L^{\mathrm{vv}}} X_{\mathrm{v}} + \overset{(sv)}{\boldsymbol{L}^{\mathrm{vq}}} \cdot \boldsymbol{X}_{\mathrm{q}} + \sum_{k=1}^{K-1} \overset{(sv)}{\boldsymbol{L}_k^{\mathrm{vd}}} \cdot \boldsymbol{X}_k' + \overset{(s\boldsymbol{a})}{\boldsymbol{L}^{\mathrm{vv}}} \cdot \boldsymbol{X}_{\mathrm{v}}^{\mathrm{a}} + \overset{(s\mathbf{t})}{\mathbf{L}^{\mathrm{vv}}} : \overset{0}{\mathbf{X}_{\mathrm{v}}^{\mathrm{s}}},$$

$$\boldsymbol{J}_{\mathrm{q}} = \boldsymbol{L} \sum_{r=1}^{R} \overset{(vs)}{{}_r^{\mathrm{qc}}} A_r + \overset{(vs)}{\boldsymbol{L}^{\mathrm{qv}}} X_{\mathrm{v}} + \overset{(vv)}{\mathbf{L}^{\mathrm{qq}}} \cdot \boldsymbol{X}_{\mathrm{q}} + \sum_{k=1}^{K-1} \overset{(vv)}{\mathbf{L}_k^{\mathrm{qd}}} \cdot \boldsymbol{X}_k' + \overset{(v\boldsymbol{a})}{\mathbf{L}^{\mathrm{qv}}} \cdot \boldsymbol{X}_{\mathrm{v}}^{\mathrm{a}} + \overset{(v\mathbf{t})}{\mathbf{L}^{\mathrm{qv}}} : \overset{0}{\mathbf{X}_{\mathrm{v}}^{\mathrm{s}}},$$

$$(3.88)$$

$$\boldsymbol{J}_i = \sum_{r=1}^{R} \overset{(vs)}{\boldsymbol{L}_{ir}^{\mathrm{dc}}} A_r + \overset{(vs)}{\boldsymbol{L}_i^{\mathrm{dv}}} X_{\mathrm{v}} + \overset{(vv)}{\mathbf{L}_i^{\mathrm{dq}}} \cdot \boldsymbol{X}_{\mathrm{q}} + \sum_{k=1}^{K-1} \overset{(vv)}{\mathbf{L}_{ik}^{\mathrm{dd}}} \cdot \boldsymbol{X}_k' + \overset{(v\boldsymbol{a})}{\mathbf{L}_i^{\mathrm{dv}}} \cdot \boldsymbol{X}_{\mathrm{v}}^{\mathrm{a}} + \overset{(v\mathbf{t})}{\mathbf{L}_i^{\mathrm{dv}}} : \overset{0}{\mathbf{X}_{\mathrm{v}}^{\mathrm{s}}},$$

$$(i = 1, 2, \ldots, K-1)$$

$$\boldsymbol{P}^{\mathrm{va}} = \sum_{r=1}^{R} \overset{(\boldsymbol{a}s)}{\boldsymbol{L}_r^{\mathrm{vc}}} A_r + \overset{(\boldsymbol{a}s)}{\boldsymbol{L}^{\mathrm{vv}}} X_{\mathrm{v}} + \overset{(\boldsymbol{a}v)}{\mathbf{L}^{\mathrm{vq}}} \cdot \boldsymbol{X}_{\mathrm{q}} + \sum_{k=1}^{K-1} \overset{(\boldsymbol{a}v)}{\mathbf{L}_k^{\mathrm{vd}}} \cdot \boldsymbol{X}_k' + \overset{(\boldsymbol{aa})}{\mathbf{L}^{\mathrm{vv}}} \cdot \boldsymbol{X}_{\mathrm{v}}^{\mathrm{a}} + \overset{(\boldsymbol{a}\mathbf{t})}{\mathbf{L}^{\mathrm{vv}}} : \overset{0}{\mathbf{X}_{\mathrm{v}}^{\mathrm{s}}},$$

$$\overset{0}{\mathbf{P}^{\mathrm{vs}}} = \sum_{r=1}^{R} \overset{(\mathbf{t}s)}{\mathbf{L}_r^{\mathrm{vc}}} A_r + \overset{(\mathbf{t}s)}{\mathbf{L}^{\mathrm{vv}}} X_{\mathrm{v}} + \overset{(\mathbf{t}v)}{\mathbf{L}^{\mathrm{vq}}} \cdot \boldsymbol{X}_{\mathrm{q}} + \sum_{k=1}^{K-1} \overset{(\mathbf{t}v)}{\mathbf{L}_k^{\mathrm{vd}}} \cdot \boldsymbol{X}_k' + \overset{(\mathbf{t}\boldsymbol{a})}{\mathbf{L}^{\mathrm{vv}}} \cdot \boldsymbol{X}_{\mathrm{v}}^{\mathrm{a}} + \overset{(\mathbf{tt})}{\mathbf{L}^{\mathrm{vv}}} : \overset{0}{\mathbf{X}_{\mathrm{v}}^{\mathrm{s}}}.$$

In this tensor form of the linear kinematical equations the upper (bracketed) indices of the conductivity tensors indicate the tensor type of the coupled fluxes and forces, whereas the side indices (not bracketed) signify the physical nature of the coupling. For example, the conductivity tensor $\overset{(sv)}{\boldsymbol{L}^{\mathrm{vq}}}$ giving account of the cross-effect of the heat conduction (q) with the viscous current density (v) which couples, according to the (sv) bracketed pair of indices, a scalar and a vector process. Hence, in general, the upper indices (ss), $(s\boldsymbol{a})$, $(\mathbf{t}\boldsymbol{a})$, $(\mathbf{tt})$, ... denote that scalar-scalar, scalar-axial vector, symmetric tensor with zero trace-axial vector, etc. type of coupling. The lower j and r indices refer to the chemical reactions, further, the i and k indices refer to the components, just as in the foregoing.

The tensorial order of the conductivity coefficients and their polar or axial character is indicated in the following table.

Tensorial order	Polar tensors	Axial tensors
0 (scalar)	$\overset{(ss)}{L}{}^{\mathrm{cc}}_{jr}, \overset{(ss)}{L}{}^{\mathrm{cv}}_{j}, \overset{(ss)}{L}{}^{\mathrm{vc}}_{r}, \overset{(ss)}{L}{}^{\mathrm{vv}}$	
1 (vector)	$\overset{(sv)}{L}{}^{\mathrm{cq}}_{j}, \overset{(sv)}{L}{}^{\mathrm{vq}}, \overset{(sv)}{L}{}^{\mathrm{cd}}_{jk}, \overset{(sv)}{L}{}^{\mathrm{vd}}_{k}, \overset{(vs)}{L}{}^{\mathrm{qc}}_{r}, \overset{(vs)}{L}{}^{\mathrm{dc}}_{ir}, \overset{(vs)}{L}{}^{\mathrm{qv}}, \overset{(vs)}{L}{}^{\mathrm{dv}}_{i}$	$\overset{(sa)}{L}{}^{\mathrm{cv}}_{j}, \overset{(sa)}{L}{}^{\mathrm{vv}}, \overset{(as)}{L}{}^{\mathrm{vc}}_{r}, \overset{(as)}{L}{}^{\mathrm{vv}}$
2 (tensor)	$\overset{(vv)}{\mathbf{L}}{}^{\mathrm{qq}}, \overset{(vv)}{\mathbf{L}}{}^{\mathrm{qd}}_{k}, \overset{(vv)}{\mathbf{L}}{}^{\mathrm{dq}}_{i}, \overset{(vv)}{\mathbf{L}}{}^{\mathrm{dd}}_{ik}, \overset{(st)}{\mathbf{L}}{}^{\mathrm{cv}}_{j}, \overset{(st)}{\mathbf{L}}{}^{\mathrm{vv}}, \overset{(ts)}{\mathbf{L}}{}^{\mathrm{vc}}_{r}, \overset{(ts)}{\mathbf{L}}{}^{\mathrm{vv}}, \overset{(aa)}{\mathbf{L}}{}^{\mathrm{vv}}$	$\overset{(va)}{\mathbf{L}}{}^{\mathrm{qv}}, \overset{(va)}{\mathbf{L}}{}^{\mathrm{dv}}_{i}, \overset{(av)}{\mathbf{L}}{}^{\mathrm{vq}}, \overset{(av)}{\mathbf{L}}{}^{\mathrm{vd}}_{k},$
3 (tensor)	$\overset{(vt)}{\mathbf{L}}{}^{\mathrm{qv}}, \overset{(vt)}{\mathbf{L}}{}^{\mathrm{dv}}_{i}, \overset{(tv)}{\mathbf{L}}{}^{\mathrm{vq}}, \overset{(tv)}{\mathbf{L}}{}^{\mathrm{vd}}_{k}$	$\overset{(at)}{\mathbf{L}}{}^{\mathrm{vv}}, \overset{(ta)}{\mathbf{L}}{}^{\mathrm{vv}}$
4 (tensor)	$\overset{(tt)}{\mathbf{L}}{}^{\mathrm{vv}}$	

b) Curie Principle. The general content of the principle is that the possible spatial symmetries of an anisotropic system reduce the coefficients occuring in the linear laws in such a manner that the Cartesian components of the currents do not all depend on all the components of the forces [36]. The principle gets a particular sense in the case of isotropic systems and the theorem which can be formulated for this particular case is the so-called Curie principle. As it has been demonstrated exactly by de GROOT and MAZUR [3], the Cartesian components of thermodynamic forces of different tensorial order and character are transformed in an isotropic system under rotation and inversion, in such a manner that it is only between the currents and forces of identical tensorial order where there remains any coupling. This can also be expressed in such a way that with the application of linear laws for the case of isotropic systems, all the tensors occuring in these are reduced to a scalar multiplied by the unit tensor, hence

$$\mathbf{L} = L\,\boldsymbol{\delta}, \tag{3.89}$$

where L is the corresponding scalar. Thus for isotropic systems the Curie principle can be formulated in the following form: *in isotropic systems phenomena which are described by thermodynamic forces and currents of different tensorial order and character* (at least in the case of interactions which can be described with linear constitutive equations) *do not interfere with one another.*

c) Isotropic Case. The systematic application of the Curie principle to the linear kinematical equations (3.88) leads to the result that in the case of isotropic continua the linear constitutive equations

$$J_j = \sum_{r=1}^{R} L_{jr} A_r + L_j^{\mathrm{cv}} X_{\mathrm{v}}, \qquad (j = 1, 2, \ldots, R) \tag{3.90}$$

$$p^{\mathrm{v}} = \sum_{r=1}^{R} L_r^{\mathrm{vc}} A_r + \overset{(ss)}{L} X_{\mathrm{v}}, \tag{3.91}$$

$$\boldsymbol{J}_{\mathrm{q}} = L_{\mathrm{qq}} \boldsymbol{X}_{\mathrm{q}} + \sum_{k=1}^{K-1} L_{\mathrm{q}k} \boldsymbol{X}_k', \tag{3.92}$$

$$\boldsymbol{J}_i = L_{i\mathrm{q}} \boldsymbol{X}_{\mathrm{q}} + \sum_{k=1}^{K-1} L_{ik} \boldsymbol{X}_k', \qquad (i = 1, 2, \ldots, K-1) \tag{3.93}$$

$$\boldsymbol{P}^{\mathrm{va}} = \overset{(aa)}{L} \boldsymbol{X}_{\mathrm{v}}^{\mathrm{a}}, \tag{3.94}$$

$$\overset{0}{\mathbf{P}}{}^{\mathrm{vs}} = \overset{(tt)}{L} \overset{0}{\mathbf{X}}{}_{\mathrm{v}}^{\mathrm{s}}, \tag{3.95}$$

have to be used. These contain already scalar coefficients only. Here with all the coefficients, which can unambiguously be differentiated, the side indices (indicating the physical nature of the coefficients) were omitted [with the exception of the indices (qq), referring to the heat conduction, which were put in the lower position]. Similarly, owing to the scalar character of the coefficients, the upper (bracketed) indices referring to the tensorial character were omitted too, with the exception of the coefficients $\overset{(ss)}{L}$, $\overset{(aa)}{L}$ and $\overset{(tt)}{L}$ of the viscous laws.

It is shown by the linear constitutive equations (3.90) to (3.95) reduced to the isotropic case that in isotropic continua any interaction occurs between irreversible processes of identical tensorial order and character only. Though the meaning of the coefficients given in (3.90) to (3.95) is evident, particularly if the actual forms of the forces defined in (3.70) to (3.75) are taken into consideration, the meaning of the coefficients L_{qq} and $\overset{(ss)}{L}$, $\overset{(aa)}{L}$, $\overset{(tt)}{L}$ related to the pure heat conduction and viscous phenomena are discussed. Since, according to the Fourier law of heat conduction

$$\boldsymbol{J}_{\mathrm{q}} = -\lambda \nabla T \tag{3.96}$$

where λ is the heat conductivity coefficient, it is shown by the comparison of (3.92) to (3.96) given in the Fourier picture that the Onsager coefficient L_{qq} of the entropy picture appears in the relation

$$L_{\mathrm{qq}} = T^2 \lambda \tag{3.97}$$

with the ordinary heat conductivity coefficient. Similarly, the comparison of Newton's linear laws of viscosity to Onsager's linear constitutive equations (3.91), (3.94) and (3.95), referring also to viscosity, results in the following identity:

$$\overset{(ss)}{L} = T\eta_{\mathrm{v}}, \quad \overset{(tt)}{L} = 2T\eta, \quad \overset{(aa)}{L} = T\eta_{\mathrm{r}} \tag{3.98}$$

where η_{v} is the volume, η the shear and η_{r} the rotational viscosity coefficient. If the linear kinematical constitutive equations belonging to the energy picture (3.85) are used instead of (3.90) to (3.95), according to (3.86) one has to divide the Onsager coefficients L_{qq} etc., given in entropy picture, by T. Thus for example

$$L^*_{\mathrm{qq}} = T^{-1}L_{\mathrm{qq}} = T\lambda \tag{3.99}$$

is Onsager's coefficient which belongs to the pure heat conduction in the energy picture.

The reduced linear relations (3.90) to (3.95) following from the Curie principle have also a further important consequence. Namely, owing to the Curie principle, there is no connection between the thermodynamic forces and currents of different tensorial order and character in the isotropic case and therefore the expression of the entropy production (3.87) is decomposable to four parts of different character, which according to the second law must be separately positive definite quantities. Hence the entropy production of the scalar forces and fluxes:

$$\overset{(s)}{\sigma} = \sum_{j=1}^{R} J_j A_j + p^{\mathrm{v}} X_{\mathrm{v}} \geqq 0, \tag{3.100}$$

the entropy production of the polar vector currents and forces:

$$\overset{(v)}{\sigma} = \boldsymbol{J}_{\mathrm{q}} \cdot \boldsymbol{X}_{\mathrm{q}} + \sum_{k=1}^{K-1} \boldsymbol{J}_k \cdot \boldsymbol{X}'_k \geqq 0, \tag{3.101}$$

the entropy production of the axial vector currents and forces:

$$\overset{(a)}{\sigma} = \boldsymbol{P}^{\mathrm{va}} \cdot \boldsymbol{X}^{\mathrm{a}}_{\mathrm{v}} \geqq 0, \tag{3.102}$$

and finally the viscous entropy production determined by symmetric tensors with zero trace

$$\overset{(t)}{\sigma} = \overset{0}{\mathbf{P}}{}^{\mathrm{vs}} : \overset{0}{\mathbf{X}}{}^{\mathrm{s}}_{\mathrm{v}} \geqq 0 \tag{3.103}$$

must separately be positive definite ones. In other words by the second law and at the same time by the Curie principle it is required that not only the condition

$$\sigma = \overset{(s)}{\sigma} + \overset{(v)}{\sigma} + \overset{(a)}{\sigma} + \overset{(t)}{\sigma} \geqq 0 \tag{3.104}$$

must be fulfilled, but that the positive definity stipulated by the second law for the entropy production should separately be fulfilled for processes, which do not interact with each other as a consequence of the Curie principle.

Let us enlighten the foregoing by an important reaction kinetic example. Let us consider the entropy production σ_c determined by (3.100) without the viscous term which can often be neglected in practical cases. If in a system only a single chemical reaction takes place, the reaction rate and the affinity must always have identical sign due to the positive definity

$$\sigma_c = JA \geqq 0 \tag{3.105}$$

valid according to the second law.

The situation is different in the case of simultaneous reaction systems. In such cases the condition

$$\sigma_c = \sum_{j=1}^{R} J_j A_j \geqq 0 \tag{3.106}$$

can be fulfilled in spite of the fact that for one group of the simultaneous reactions, for example

$$\overset{(1)}{\sigma_c} = \sum_{j=1}^{r} J_j A_j \leqq 0 \tag{3.107}$$

is valid, if at the same time for the group

$$\overset{(2)}{\sigma_c} = \sum_{j=r+1}^{R} J_j A_j \geqq 0 \tag{3.108}$$

the positive definity is fulfilled to such an extent that the entropy production of the whole reacting system σ_c is positive definite according to (3.106). However, this is possible only if the condition

$$\overset{(2)}{\sigma_c} \geqq \left|\overset{(1)}{\sigma_c}\right| \tag{3.109}$$

is valid, which for the case of two simultaneous reactions can be written as

$$J_2 A_2 \geqq |J_1 A_1| . \tag{3.110}$$

However, it is evident that this condition allows the following possibilities:

$$J_1 A_1 \leqq 0 \tag{3.111a}$$

and

$$J_2 A_2 \geqq 0 \tag{3.111b}$$

where, concerning the first formula it can be stated that the reaction proceeds in a direction opposite to affinity. This case is possible owing to the positive definity of the total entropy production only if the entropy production produced by the second simultaneous reaction supplies the entropy decrease of the reaction proceeding in the direction opposite to its own affinity.

What has been said concerning chemical reactions can be generalized—mutatis mutandis—for the case of other irreversible processes as well. Thus, for example, if heat conduction and diffusion take place simultaneously in a system, (3.101) must be fulfilled according to the second law. Of course, it is also possible that the heat flow proceeds in the direction of the increasing temperature gradient if the decrease in entropy connected with this process is replaced by the entropy production caused by diffusion. Though these and other similar cases are of great importance both in practical and theoretical respects we have to dispense with further examination. Anyway, after classification of the tensorial order and character of irreversible processes the Curie principle helps to apply the second law ensuring the positive definity of the entropy production for the entire process in a more detailed, hence in a more sophisticated form as in the original case.

5. Reciprocal Relations

With the aid of the Curie principle, information is obtained on the reduction of the number of the independent coefficients occuring in the linear constitutive equations by the spatial symmetry of matter. The Onsager-Casimir reciprocal relations, which are concluded from the property of time reversal invariance of the equations of motion of the individual particles—of the classical as well as of the quantum mechanical ones—lead to a further reduction of the coefficients of linear laws. The relations in question have been originally derived by Onsager for the case of the so-called "α"-type variables, i.e. for ones, which are even functions of the velocities of the molecular particles [22]. Later Casimir modified and extended the validity of the reciprocal relation also for the case of the so-called "β"-parameters, i.e. for ones which are odd functions of the particle velocities [37]. The general form of the Onsager-Casimir reciprocal relations containing both cases (in a scalar form and in the case of zero magnetic field and angular velocity) are

$$L_{ik} = \varepsilon_i \varepsilon_k L_{ki}, \qquad (i, k = 1, 2, \ldots, f) \tag{3.112}$$

where f is the number of the independent scalar fluxes and forces occuring in the linear kinematical constitutive equations. Here $\varepsilon_i = \varepsilon_k = 1$, if the coefficient L_{ik} refers to cross-effects which can be described either purely by α or purely by β parameters, whereas $\varepsilon_i = 1$ and

$\varepsilon_k = -1$ in the mixed case. Therefore, according to the cases

$$\varepsilon_i \varepsilon_k = \begin{cases} 1 & \text{Onsager,} \\ -1 & \text{Casimir,} \end{cases}$$

we speak of Onsager's symmetric and Casimir's antisymmetric reciprocal relations, respectively. Assuming that between f independent scalar parameters there are $1, 2, \ldots, m$; α type and $m+1, \ldots, f$; β type parameters, the general Onsager's and Casimir's reciprocal relations (in the case of zero magnetic field and angular velocity) are the following:

$$\begin{aligned} &\text{Onsager } L_{ik} = L_{ki}, && (i, k = 1, 2, \ldots, m) \\ &\text{Casimir } L_{i\nu} = -L_{\nu i}, && (i = 1, 2, \ldots, m; \nu = m+1, \ldots, f) \\ &\text{Onsager } L_{\lambda\nu} = L_{\nu\lambda}, && (\lambda, \nu = m+1, \ldots, f). \end{aligned} \tag{3.113}$$

Here we cannot give the verification of these reciprocal relations. However, it should be mentioned that their validity relies upon the hypothesis of microscopic reversibility and usually these are verified by making use of the fluctuation theory. (See, for instance [3, 4] and further, the original works [22, 37]). Hence the validity of the reciprocal relation is based—at least hitherto—upon a hypothesis of mechanical origin, which is alien from thermodynamics. On the other hand, it is also certain that the fluctuation-theoretical proofs using the mathematical apparatus of the theory of stochastic processes cannot be considered a true statistical derivation, at least not in the exact Gibbsian sense.[1] Therefore, in phenomenological thermodynamics, if we intend to remain consequent to the spirit of this theory, the reciprocal relations must either be considered experimentally confirmed axioms or one must endeavour their direct phenomenological derivation. Though many attempts were made in the latter direction in the past years [31, 38, 39, 40, 41], further efforts are necessary. Indeed the opinion of TRUESDELL, viz. "If the reciprocal relations are true, it must be possible to derive them by pure phenomenological means also" [42], is absolutely right.

With the aid of the general reciprocal relations (3.113), the adequate reciprocal relations for the coefficients of the linear constitutive equations containing the independent fluxes and forces in actual cases can immediately be given.

a) Anisotropic Case. Since the forces and fluxes of the linear relations (3.88) are independent, the corresponding reciprocal relations for the coefficients can be written directly. However, it must be taken into account that in (3.88) the viscous forces X_{v}, $\boldsymbol{X}_{\mathrm{v}}^{\mathrm{a}}$ and $\overset{0}{\mathbf{X}}{}_{\mathrm{v}}^{\mathrm{s}}$ are β type variables, because the velocity $\boldsymbol{v}$ is a typical β parameter. Therefore, if we consider the scheme of the coefficients of (3.88) as a hypermatrix,

[1] We have to note that recently successful research was carried out by J. P. TERLETSKY and his coworkers to show that the exact theorems of fluctuations and correlations can be obtained by general methods of statistical mechanics, that is, by the exact Gibbs' method [82, 83, 84, 85].

the diagonal blocks are symmetric:

$$
\begin{aligned}
&\overset{(ss)}{L}{}^{\mathrm{cc}}_{rj} = \overset{(ss)}{L}{}^{\mathrm{cc}}_{jr}, \qquad (r, j = 1, 2, \ldots, R) && \left\{\frac{1}{2}\, R(R-1)\right\},\\
&\overset{(vv)}{\mathbf{L}}{}^{\mathrm{qq}} = \overset{\widetilde{(vv)}}{\mathbf{L}}{}^{\mathrm{qq}}, && \{3\},\\
&\overset{(vv)}{\mathbf{L}}{}^{\mathrm{dd}}_{ik} = \overset{\widetilde{(vv)}}{\mathbf{L}}{}^{\mathrm{dd}}_{ki}, \qquad (i, k = 1, 2, \ldots, K-1), && \left\{\frac{3(K-1)\,[3(K-1)-1]}{2}\right\},\\
&\overset{(aa)}{\mathbf{L}}{}^{\mathrm{vv}} = \overset{\widetilde{(aa)}}{\mathbf{L}}{}^{\mathrm{vv}}, && \{3\},\\
&\overset{(tt)}{\mathbf{L}}{}^{\mathrm{vv}} = \overset{\widetilde{(tt)}}{\mathbf{L}}{}^{\mathrm{vv}}, && \{10\}.
\end{aligned}
\tag{3.114}
$$

Here the transposed tensors with respect to space coordinates have been denoted by $\sim$, and in the right-hand side bracket $\{\}$ the number of the reciprocal relations was given with respect to the scalar coefficients. In the same sense, the off-diagonal block consisting of the scalar elements of tensors of different order are transposed

$$
\begin{aligned}
&\overset{(vs)}{\boldsymbol{L}}{}^{\mathrm{qc}}_{j} = \overset{(sv)}{\boldsymbol{L}}{}^{\mathrm{cq}}_{j}, \qquad (j = 1, 2, \ldots, R) && \{3R\},\\
&\overset{(vs)}{\boldsymbol{L}}{}^{\mathrm{dc}}_{ij} = \overset{(sv)}{\boldsymbol{L}}{}^{\mathrm{cd}}_{jc}, \qquad (j = 1, 2, \ldots, R;\ i = 1, 2, \ldots, K-1) && \{3R(K-1)\},\\
&\overset{(vv)}{\mathbf{L}}{}^{\mathrm{dq}}_{i} = \overset{\widetilde{(vv)}}{\mathbf{L}}{}^{\mathrm{qd}}_{i}, \qquad (i = 1, 2, \ldots, K-1), && \{9(K-1)\},\\
&\overset{(as)}{\boldsymbol{L}}{}^{\mathrm{vv}} = \overset{(sa)}{\boldsymbol{L}}{}^{\mathrm{vv}}, && \{3\},\\
&\overset{(ts)}{\mathbf{L}}{}^{\mathrm{vv}} = \overset{\widetilde{(st)}}{\mathbf{L}}{}^{\mathrm{vv}}, && \{5\},\\
&\overset{(ta)}{\mathbf{L}}{}^{\mathrm{vv}} = \overset{\widetilde{(at)}}{\mathbf{L}}{}^{\mathrm{vv}}, && \{15\},
\end{aligned}
\tag{3.115}
$$

and negative transposed of each other, respectively:

$$
\begin{aligned}
&\overset{(ss)}{L}{}^{\mathrm{vc}}_{j} = -\overset{(ss)}{L}{}^{\mathrm{cv}}_{j}, \qquad (j = 1, 2, \ldots, R), && \{R\},\\
&\overset{(vs)}{\boldsymbol{L}}{}^{\mathrm{qv}} = -\overset{(sv)}{\boldsymbol{L}}{}^{\mathrm{vq}}, && \{3\},\\
&\overset{(vs)}{\boldsymbol{L}}{}^{\mathrm{dv}}_{i} = -\overset{(sv)}{\boldsymbol{L}}{}^{\mathrm{vd}}_{i}, \qquad (i = 1, 2, \ldots, K-1), && \{3(K-1)\},\\
&\overset{(as)}{\boldsymbol{L}}{}^{\mathrm{vc}}_{j} = -\overset{(sa)}{\boldsymbol{L}}{}^{\mathrm{cv}}_{j}, \qquad (j = 1, 2, \ldots, R), && \{3R\},
\end{aligned}
$$

$$\overset{(ts)}{\mathbf{L}}{}_j^{\mathrm{vc}} = -\overset{\widetilde{(st)}}{\mathbf{L}}{}_j^{\mathrm{cv}}, \qquad (j = 1, 2, \ldots, R), \qquad \{5R\}, \quad (3.116)$$

$$\overset{(av)}{\mathbf{L}}{}^{\mathrm{vq}} = -\overset{\widetilde{(va)}}{\mathbf{L}}{}^{\mathrm{qv}}, \qquad \{9\},$$

$$\overset{(tv)}{\mathbf{L}}{}^{\mathrm{vq}} = -\overset{\widetilde{(vt)}}{\mathbf{L}}{}^{\mathrm{qv}}, \qquad \{15\},$$

$$\overset{(av)}{\mathbf{L}}{}_i^{\mathrm{vd}} = -\overset{\widetilde{(va)}}{\mathbf{L}}{}_i^{\mathrm{dv}}, \qquad (i = 1, 2, \ldots, K-1), \qquad \{9(K-1)\},$$

$$\overset{(tv)}{\mathbf{L}}{}_i^{\mathrm{vd}} = -\overset{\widetilde{(vt)}}{\mathbf{L}}{}_i^{\mathrm{dv}}, \qquad (i = 1, 2, \ldots, K-1), \qquad \{15(K-1\}.$$

Relations (3.114) and (3.115) are of the Onsager type, whereas those of (3.116) are Casimir's reciprocal relations. We note that in (3.114) the last two Onsager relations, as well as the last three Onsager relations of (3.115) are the explicit forms of the last relations of the general relations (3.113). Hence, they follow from the relation valid for the coupling between purely β parameters.

b) Isotropic Case. Now let us apply the general Onsager-Casimir relations given in (3.113) for the scalar coefficients of the linear constitutive equations (3.90) to (3.95). For the scalar coefficients in question it can directly be written that

$$\begin{aligned} &L_{jr} = L_{rj}, && (r, j = 1, 2, \ldots, R), && \left\{\frac{1}{2} R(R-1)\right\}, \\ &L_j^{\mathrm{vc}} = -L_j^{\mathrm{cv}}, && (j = 1, 2, \ldots, R), && \{R\}, \\ &L_{i\mathrm{q}} = L_{\mathrm{q}i}, && (i = 1, 2, \ldots, K-1), && \{K-1\}, \\ &L_{ik} = L_{ki}, && (i, k = 1, 2, \ldots, K-1), && \left\{\frac{1}{2}(K-1)(K-2)\right\}. \end{aligned} \qquad (3.117)$$

By the reciprocal relations it is shown that the coefficients describing the cross-effects between the chemical reactions and the volume viscosity fulfil Casimir's antisymmetry, because the A_j affinities are α type forces, whereas the viscous force $X_{\mathrm{v}} \equiv -\frac{\nabla \cdot \boldsymbol{v}}{T}$ is a β type one. It is evident that these reciprocal relations are obtained directly also in the case, when the Curie principle is used for the linear kinematical constitutive equations (3.88) by considering the reciprocal relations (3.114), (3.115) and (3.116).

Second Part

Variational Principles

On the Variational Principles in General

It is wellknown that in some branches of physics of a high degree of exactitude, the deductive mathematical methods are becoming increasingly important besides the genetic and inductive research methods. In other words, besides the application of the genetic axiomatic method, the constructive methods are also accepted in teaching, in development and in the practical applications of the theory. Hence, the essential part of the constructive axiomatic method is seen in the deductive development of a physical discipline which is mainly carried out with variational calculus.

The variational calculus is an incomparable all-embracing mathematical method. With its aid the details of the topics in question can be reproduced and their fundamental equations can be derived from the variational principles expressing the exact physical laws completely. Owing to the fact that with the variational technique, the development of several branches of physics was possible the variational calculus can be regarded as an important developing method, which has a heuristical value as well. Also taking into consideration that instead of solving the differential equations describing the different problems in physical and technical sciences by formulating the task in a variational form, the solution is arrived at by applying one of the direct methods of the variational calculus, the direct practical advantages of the variational computation need not be stressed any further.

The most characteristical features concerning the general nature of the variational principles are the following. A variational principle always contains:

1. *Statements referring to the model of systems as whole.*

2. *It refers to the extremum of a scalar function and thus ensuring an a priori invariant description.*

3. *It always includes the differential equations representing the fundamental equations of the branches of sciences in question together with the boundary, jump and constraint conditions.*

The general properties of the variational principles, their historical and gnosiological rôle cannot be dealt with in the present work. (In this respect see: [43, 44, 45].)

IV. The Principle of Least Dissipation of Energy

In a way similar to the derivation of the fundamental equations of mechanics and electrodynamics from the wellknown variational principles (d'Alembert principle, Gauss' principle of the least constraint, which are differential principles) or from variational principles as understood in a closer sense (Maupertuis' principle of the least action and particularly the Hamilton principle, which are integral principles), the fundamental equations of thermodynamics can be all-embraced with a single variational principle as well. This principle was originally formulated by Onsager and was called: *the principle of least dissipation of energy* [22]. The first formulation of this principle was restricted to the particular case of heat conduction in anisotropic continua, and no considerable generalization was attained, when in 1953 Onsager and Machlup [46], and after then in 1957 Tisza and Manning [47] extended the validity of the principle for the case of adiabatically isolated, noncontinuous systems. Formulation of the principle for the above-mentioned particular cases meant a limitation, by which up to now the general development and widespread practical application of the principle was hindered. This is shown by the fact that in most detailed monographs the principle of least dissipation of energy is not even mentioned [3, 8, 13, 17, 25, 27] with the exception of [4], where, however, less is found as in the original works.

Important viewpoints were raised by Ono in 1961 [48], even if earlier but unpublished results are also taken into account [31]. By that time, *the principle of minimum production of entropy* recognized by Prigogine [49, 34, 17, 23] was already applied on a large scale, but the relation of this principle to Onsager's principle was not analysed. Ono was the first who made attempts to clarify the relation between the principle of least dissipation of energy and the principle of the minimum entropy production. He was in a fairly difficult position, since the particular formulations of the Onsager principle known until that time gave no possibility to a general analysis. Since the main feature of the particular formulations was that variation is carried out according to the fluxes in case of constant forces, Ono was right in considering the difference between the Onsager and the Prigogine principles in the convention of variation, since the basic feature of the Prigogine principle is to vary simultaneously with respect to the fluxes and forces. Ono's conclusions provocated a vivid reaction (GYARMATI [50, 51, 52, 53], KIRKALDY [54]). Notably, an alternative representation of the Onsager principle, where one has to vary with respect to the forces in case of constant fluxes was already known earlier (GYARMATI [31]). Though the final clarification of the relation between Onsager's and Prigogine's principle was finished only recently [55], the above-mentioned examinations contributed to the general development and extension of the range of application of the Onsager principle, moreover also to the recognition of a new integral principle of thermodynamics.

With the following development of the principle of least dissipation of energy we rely upon the works [50, 51, 52, 53]. Onsager's original formulations [22, 46], given for particular cases, are derived from the new and general forms of the principle. It will be seen that the general formulation of the principle corresponds in every respect to the thinking of field theory. Hence, the equations of the latter are also incorporated in the principle and thus it can be used to examine all processes discussed in preceding chapters.

1. Non-equilibrium Potential Functions

Let us consider the bilinear expression of the entropy production

$$\sigma(\boldsymbol{J}, \boldsymbol{X}) = \sum_{i=1}^{f} J_i X_i \geqq 0, \tag{4.1}$$

where f is the number of all independent scalar fluxes J_i and forces X_i. Since, the splitting of the entropy production into the sum of the corresponding number of products of fluxes and conjugated forces can be carried out in every case, (4.1) is valid quite generally, and independent of the fact whether linear relations between the fluxes and forces have been given or not. In the case of linear constitutive equations

$$J_i = \sum_{k=1}^{j} L_{ik} X_k, \qquad (i = 1, 2, \ldots, f) \tag{4.2}$$

and due to Onsager's reciprocal relations valid for the coefficients

$$L_{ik} = L_{ki}, \qquad (i, k = 1, 2, \ldots, f) \tag{4.3}$$

σ can also be given as the homogeneous quadratic expression of the independent thermodynamic forces, i.e.

$$\sigma = \sum_{i,k=1}^{f} L_{ik} X_i X_k \geqq 0. \tag{4.4}$$

In contrast to (4.1) this expression is already based on the validity of the linear constitutive equations (4.2). Since, at least for irreversible transformations, σ must be positive definite according to the second law, this implies that all diagonal elements of the symmetric matrix $L_{ik} + L_{ki}$ are positive, whereas the off-diagonal elements must satisfy the following conditions: $L_{ii}L_{kk} \geqq \frac{1}{4}(L_{ik} + L_{ki})$.

Sometimes alternative forms of the expressions (4.2), (4.3) and (4.4), which are expressed in terms of the resistances R_{ik} instead of the conductivity coefficients L_{ik} are used. By introducing the resistance matrix R_{ik} as the reciprocal of the matrix L_{ik} we have

$$\sum_{m=1}^{f} L_{im}R_{mk} = \sum_{m=1}^{f} R_{im}L_{mk} = \delta_{ik}, \qquad (i, k = 1, 2, \ldots, f) \tag{4.5}$$

where δ_{ik} is the Kronecker symbol, and in this case the alternative forms of the expressions (4.2), (4.3) and (4.4) are obtained:

$$X_i = \sum_{k=1}^{f} R_{ik}J_k, \qquad (i = 1, 2, \ldots, f) \tag{4.6}$$

$$R_{ik} = R_{ki}, \qquad (i, k = 1, 2, \ldots, f) \tag{4.7}$$

$$\sigma = \sum_{i,k=1}^{f} R_{ik}J_iJ_k \geqq 0. \tag{4.8}$$

The expressions (4.2), (4.3) and (4.4) and their alternative forms (4.6), (4.7) and (4.8) are equivalent to one another. This fact is the criterion of subsequent considerations.

The expressions (4.4) and (4.8) given for σ can be considered as non-equilibrium potential functions due to the validity of the symmetry relations (4.3) and (4.7). However, other non-equilibrium potential functions can be directly defined by the so-called *dissipation functions*, introduced originally in particular cases by RAYLEIGH [56] and ONSAGER [22, 46]. We now define the local correspondents of these functions quite generally, by the following homogeneous quadratic forms:

$$\Psi(\boldsymbol{X}, \boldsymbol{X}) \equiv \frac{1}{2} \sum_{i,k=1}^{f} L_{ik}X_iX_k \geqq 0 \tag{4.9}$$

and

$$\Phi(\boldsymbol{J}, \boldsymbol{J}) \equiv \frac{1}{2} \sum_{i,k=1}^{f} R_{ik}J_iJ_k \geqq 0. \tag{4.10}$$

These functions may be called more precise *local dissipation potentials* and they are in the case of the validity of linear constitutive equations equal to half of the entropy production. Hence, similar to σ, Ψ and Φ are also the local measures of the irreversibility, and differ from one another only in the way of the description of the non-equilibrium state. Namely $\Psi(\boldsymbol{X}, \boldsymbol{X})$ depends on the forces wich determine the non-equilibrium state itself, whereas $\Phi(\boldsymbol{J}, \boldsymbol{J})$ is the function of the fluxes (general velocities or rates) characterizing the state variations.

The potential character of the functions Ψ and Φ for the fluxes and forces respectively, can be seen from the linear constitutive equations

$$J_i = \frac{\partial \Psi}{\partial X_i} = \sum_{k=1}^{f} L_{ik} X_k, \qquad (i = 1, 2, \dots, f) \tag{4.11}$$

and

$$X_i = \frac{\partial \Phi}{\partial J_i} = \sum_{k=1}^{f} R_{ik} J_k, \qquad (i = 1, 2, \dots, f) \tag{4.12}$$

following from (4.9) and (4.10), and further on the basis of the relations

$$\frac{\partial^2 \Psi}{\partial X_i \, \partial X_k} = \frac{\partial J_i}{\partial X_k} = L_{ik} = L_{ki} = \frac{\partial J_k}{\partial X_i} = \frac{\partial^2 \Psi}{\partial X_k \, \partial X_i}, \qquad (i, k = 1, 2, \dots, f) \tag{4.13}$$

and

$$\frac{\partial^2 \Phi}{\partial J_i \, \partial J_k} = \frac{\partial X_i}{\partial J_k} = R_{ik} = R_{ki} = \frac{\partial X_k}{\partial J_i} = \frac{\partial^2 \Phi}{\partial J_k \, \partial J_i}, \qquad (i, k = 1, 2, \dots, f) \tag{4.14}$$

which express the reciprocal relations. Hence the first derivatives of the dissipation functions contain the linear constitutive equations, whereas the equality of the second mixed derivatives are equivalent to the existence of the Onsager reciprocal relations. Accordingly, the potentials Ψ and Φ are functions which were constructed advisable for the linear Onsager theory, and in which the premises of the Onsager theory are accumulated, i.e. the linear constitutive equations together with the reciprocal relations valid for the coefficients. On the other hand, it is evident from (4.13) and (4.14) that the existence of the reciprocal relations means that the f dimension "abstract spaces" of the force parameters $\{X_1, X_2, \dots, X_f\}$ and of the fluxes $\{J_1, J_2, \dots, J_f\}$ are "rotation-free" in the linear Onsager theory, and thus Ψ and Φ are indeed the potential functions of the correspondent irrotational "abstract spaces".

Sometimes it is preferable instead of the functions Ψ and Φ of the dimension of th entropy production, to use the following dissipation potentials

$$\Psi^* \equiv T\Psi = \frac{1}{2} \sum_{i,k=1}^{f} L_{ik}^* X_i^* X_k^* \geqq 0 \tag{4.15a}$$

and

$$\Phi^* \equiv T\Phi = \frac{1}{2} \sum_{i,k=1}^{f} R_{ik}^* J_i J_k \geqq 0 \tag{4.15b}$$

belonging directly to the energy dissipation $T\sigma$. It is shown by the applications that in the cases of non-isothermal problems the use of the dissipation potentials given in the entropy picture is more advantageous, whereas in the case of isothermal problems it is preferable to use the energy picture.

2. The Local Forms of the Principle

The principle of least dissipation of energy will be primarily formulated in local forms [50, 51, 52]. The method corresponds to the spirit of the field theories and it will be seen that the entire theory of non-equilibrium thermodynamics can be summarized with the variational principle of the least dissipation of energy. First of all the flux representation of the principle will be treated as proposed by ONSAGER for the non-local particular cases mentioned already. It should be emphasized that in local form the variational principle was not given by ONSAGER, not even for particular cases [22, 46]. The global formulation of the principle will be dealt with later on, after the description of all local representations.

a) The Flux Representation. The flux representation of the principle can be obtained, if the expression

$$[\varrho\dot{s} + \nabla \cdot \boldsymbol{J}_s - \Phi] = [\sigma(\boldsymbol{J}, \boldsymbol{X}) - \Phi(\boldsymbol{J}, \boldsymbol{J})] \tag{4.16}$$

which can be given with the substantial entropy balance (3.16), is varied with respect to the fluxes J_i in case of constant forces X_i. In (4.16), of course, the entropy production σ is considered according to (4.1) as a function of the fluxes J_i and the forces X_i. Whereas the dissipation potential is taken into account as a function determined only by the fluxes J_i in a homogeneous quadratic way according to (4.10). Carrying out the variation, we get

$$\begin{aligned} &\delta[\sigma(J_i, X_i) - \Phi(J_i, J_k)]_{X_i} \\ &= \sum_{k=1}^{f} \left\{ \frac{\partial}{\partial J_k} [\sigma(J_i, X_i)]_{X_i} - \frac{\partial}{\partial J_k} [\Phi(J_i, J_k)] \right\} \delta J_k . \end{aligned} \tag{4.17}$$

Since, however, from (4.1) it follows that

$$\left(\frac{\partial \sigma}{\partial J_k}\right)_{X_i} = X_k, \qquad (k = 1, 2, \ldots, f) \tag{4.18}$$

equation (4.17) can be written on the basis of (4.12) in the following form

$$\delta[\sigma - \Phi]_X = \sum_{k=1}^{f} \left(X_k - \frac{\partial \Phi}{\partial J_k}\right) \delta J_k = 0 \tag{4.19}$$

which is the necessary condition of the extremum. Since the dissipation potential Φ is according to (4.10) a homogeneous quadratic and positive definite function of the independent fluxes J_i, the extremum prescribed by the condition (4.19) can be a maximum only. It is also evident that this maximum is a direct consequence of the Carnot-Clausius theorem, i.e. of the second law. Anyway, the local variational principle formulated as

$$[\varrho\dot{s} + \nabla \cdot \boldsymbol{J}_s - \Phi]_X = [\sigma - \Phi]_X = \max \tag{4.20}$$

is equivalent to the linear constitution equations (4.12) and also contains the Onsager reciprocal relations directly in the form of (4.7). It is remarkable that in the case of the definition (4.10) of the dissipation potential Φ the extremum condition (4.19) is automatically satisfied by the linear constitutive equations (4.12). In other words, the linear relations (4.12) together with the reciprocal relations (4.7) are a priori sufficient (but at least in general are not necessary) for the fulfilment of the extremum condition (4.19).

b) The Force Representation. In addition to the flux representation the principle of least dissipation of energy can be reformulated. The alternative formulation will be called in the following *force representation,* because with such a formulation of the principle the dissipation function Ψ is used and we vary with respect to the forces in the case of constant fluxes. The possibility of the formulation in the force representation is self-explanatory, namely σ is a symmetric bilinear form in the terms of fluxes and forces, furthermore, there is no doubt that the fundamental equations (4.2), (4.3), (4.4), (4.10) of the linear Onsager theory are in principle equivalent to the alternative ones (4.6), (4.7), (4.8), (4.9). However, or even the more, the formulation of the principle in the force representation was effected only in the year 1957 [31], and the fact that it is more productive from practical viewpoints than the flux representation was recognized even later [50, 51]. The force representation of the principle is obtained, if instead of (4.16) the expression

$$[\varrho \dot{s} + \nabla \cdot \boldsymbol{J}_s - \Psi] = [\sigma(\boldsymbol{J}, \boldsymbol{X}) - \Psi(\boldsymbol{X}, \boldsymbol{X})] \tag{4.21}$$

is varied with respect to forces in case of constant fluxes:

$$\begin{aligned} &\delta[\sigma(J_i, X_i) - \Psi(X_i, X_k)]_{J_i} \\ &= \sum_{k=1}^{f} \left\{ \frac{\partial}{\partial X_k} [\sigma(J_i, X_i)]_{J_i} - \frac{\partial}{\partial X_k} [\Psi(X_i, X_k)] \right\} \delta X_k \,. \end{aligned} \tag{4.22}$$

It follows from (4.1) that

$$\left(\frac{\partial \sigma}{\partial X_k}\right)_{J_i} = J_k, \qquad (k = 1, 2, \ldots, f) \tag{4.23}$$

and consequently by making use of (4.11) the relation (4.22) leads to the extremum condition

$$\delta[\sigma - \Psi]_J = \sum_{k=1}^{f} \left(J_k - \frac{\partial \Psi}{\partial X_k}\right) \delta X_k = 0 \tag{4.24}$$

which is equivalent to the local extremum principle:

$$[\varrho \dot{s} + \nabla \cdot \boldsymbol{J}_s - \Psi]_J = [\sigma - \Psi]_J = \max. \tag{4.25}$$

This alternative form of the Onsager principle is equivalent to the linear constitutive equations (4.11) and also contains the Onsager reciprocal relations directly in the form (4.3). In other words, the linear relations (4.11) together with the reciprocal relations are a priori sufficient, but at least in general not necessary to fulfil the extremum condition (4.24).

The theoretical equivalence of the flux and force representation of the principle of least dissipation of energy is certain. The fact that from practical points of view the force representation is more productive than the flux representation, will be evident from the results of Chapter V and VI.

c) Universal Local Form of the Principle. The alternative representations of the principle of least dissipation of energy suggest the idea to unify the two representations in an universal form [50, 51, 52]. The universal local form of the principle was arrived at on the basis of a question originally raised by Prigogine [57]: whether or not in non-equilibrium thermodynamics a Gaussian type extremum principle is valid? In the following it will be demonstrated, in which sense the flux and force representation of the Onsager principle can be unified to an universal principle, and furthermore, how the universal principle can be transcribed in a form similar to the Gauss principle of mechanics.

In order to unify the two forms, let the local principle formulated in the flux and force representations be given once more together with the variational prescriptions:

$$\delta[\sigma(\boldsymbol{J}, \boldsymbol{X}) - \Phi(\boldsymbol{J}, \boldsymbol{J})]_{\boldsymbol{X}} = 0, \qquad \boldsymbol{X} = \text{const}, \; \delta\boldsymbol{X} = 0, \; \delta\boldsymbol{J} \neq 0 \tag{4.26}$$

and

$$\delta[\sigma(\boldsymbol{J}, \boldsymbol{X}) - \Psi(\boldsymbol{X}, \boldsymbol{X})]_{\boldsymbol{J}} = 0, \qquad \boldsymbol{J} = \text{const}, \; \delta\boldsymbol{J} = 0, \; \delta\boldsymbol{X} \neq 0. \tag{4.27}$$

Since in the first case we vary only with respect to the fluxes, whereas in the second case only with respect to the forces, therefore an arbitrary positive definite function depending only on the forces can always be added to or subtracted from (4.26). Similarly an arbitrary function depending on the fluxes can be added to or subtracted from (4.27). As such an additive function let us choose $\Psi(\boldsymbol{X}, \boldsymbol{X})$ in the first case, whereas in the second case $\Phi(\boldsymbol{J}, \boldsymbol{J})$. This is possible because due to the variational condition $\delta\boldsymbol{X} = 0$ stipulated for the flux representation (4.26) evidently $\delta\Psi = 0$, whereas in the force representation (4.27) owing to $\delta\boldsymbol{J} = 0$, the condition $\delta\Phi = 0$ is also automatically fulfilled. Owing to the fact that the convention of variation can be chosen arbitrary, the representations (4.26) and (4.27) can also be given in a single universal form:

$$\delta[\sigma - (\Psi + \Phi)] = 0. \tag{4.28}$$

This unified form of the variational principle after prescribing the kind of variation will be reduced to one of the representations (4.26) or (4.27). Moreover, it is also true that as we are varying in (4.28) simultaneously with respect to the fluxes and forces but independently from one another, in the local universal principle (4.28) the whole Onsager theory is contained in both representations.

The foregoing can be the most easily seen, if instead of (4.16) and (4.21) respectively, the expression

$$[\varrho\dot{s} + \nabla\cdot\boldsymbol{J}_s - (\Psi + \Phi)] = \{\sigma(\boldsymbol{J}, \boldsymbol{X}) - [\Psi(\boldsymbol{X}, \boldsymbol{X}) + \Phi(\boldsymbol{J}, \boldsymbol{J})]\} \quad (4.29)$$

is simultaneously varied over the fluxes as well as over the forces. It is evident without repeating the calculations specified earlier, that the extremum condition (4.28) leads to the local extremum principle

$$\sigma - (\Psi + \Phi) = \max. \quad (4.30)$$

Indeed, rewriting the condition (4.28) explicitly with (4.1), (4.9) and (4.10), we get

$$\delta\left[\sum_{i=1}^{f} J_i X_i - \frac{1}{2}\sum_{i,k=1}^{f} L_{ik} X_i X_k - \frac{1}{2}\sum_{i,k=1}^{f} R_{ik} J_i J_k\right] = 0 \quad (4.31)$$

with the simultaneous variation of which with respect to the fluxes and forces the relation

$$\sum_{i=1}^{f} (\delta J_i X_i + \delta X_i J_i) - \sum_{i,k=1}^{f} L_{ik} X_k\, \delta X_i - \sum_{i,k=1}^{f} R_{ik} J_k\, \delta J_i = 0,$$

i.e.

$$\sum_{i=1}^{f}\left(X_i - \sum_{k=1}^{f} R_{ik} J_k\right)\delta J_i + \sum_{i=1}^{f}\left(J_i - \sum_{k=1}^{f} L_{ik} X_k\right)\delta X_i = 0 \quad (4.32)$$

is obtained. Herewith it is clearly shown that the extremum principle (4.28) contains the linear Onsager theory in the flux as well as in the force representation. It is easily demonstrated that the extremum principle (4.28) remains also valid for non-linear problems if in these cases potential functions more general than (4.9) and (4.10) exist [51].

The foregoing must be completed by two remarks. Primarily the extremum principle (4.28) is not a true variational principle of non-equilibrium thermodynamics, but only a local differential principle, which must be fulfilled in each point of a continuum. This will become quite evident from the Gaussian form to be described in the followings, which is equivalent to (4.28). Nevertheless, the local differential principle (4.28) is equivalent to the whole Onsager theory, further, it can be widely applied also to the treatment of local constraint problems.

Our second remark refers to the fact that the original recognition [50, 51] of the local differential principle (4.28) was done independently from Onsager's and Machlup's works [46] (research work aimed at the solution of Prigogine's dilemma has lead to it). It is, however, easy to see that in (4.30) there is the question of the extremum of the local Onsager-Machlup (briefly OM) function, hence

$$\mathcal{O} \equiv \sigma - (\Psi + \Phi). \tag{4.33}$$

The global equivalent of this function has been derived by ONSAGER and MACHLUP [46] for the particular case of adiabatically isolated systems. Anyway, the concise forms of the local differential principle

$$\delta \mathcal{O} = 0 \tag{4.34}$$

respectively

$$\mathcal{O} = \max \tag{4.35}$$

give more information than the global principle formulated by ONSAGER and MACHLUP not only because the latter refers to a particular case, *but rather because Onsager and Machlup have never varied with respect to the forces!* With the universal form of the principle of least dissipation of energy we maintain that the variation over the fluxes as well as over the forces is permissible. In other words, we consider the universal extremum conditions (4.28) or (4.34) and the extremum principles (4.30) or (4.35) in every respect as universal form of the partial representations (4.26) and (4.27) including also the prescription of variation [50, 51, 52].

3. The Gaussian Form of the Local Principle

In the year 1954, on the basis of the principle of minimum production of entropy, Prigogine came to the conclusion that the Gaussian principle of least constraint is—mutatis mutandis—valid in thermodynamics too [57]. He gave no exact answer to the problem raised. However, Prigogine stated that for a diagonal form of the linear constitutive equations (4.6) (which is always obtainable by an orthogonal transformation) the minimum principle of the entropy production requires that

$$\sigma = \sum_{i=1}^{f} R_{ii} J_i^2 = \sum_{i=1}^{f} \frac{J_i^2}{L_{ii}} = \min. \tag{4.36}$$

With the assumption of identical conductivity coefficients Prigogine came to this conclusion that the principle

$$\overline{J_i^2} = \min, \tag{4.37}$$

(where an average of the square of the fluxes or velocities is considered) is similar to the Gauss principle of mechanics. However, Prigogine was aware of the fact that his assumption is a very rough one, further that the similarity of (4.37) to the Gaussian principle is a little obscure, therefore he summarized his results as follows [57]: "D'une manière un peu obscure mais imagée on pourrait appeler ce principe celui de la *moindre vitesse*".

There is no doubt that the "principle" (4.37) is unclear and practically useless. However, Prigogine's assumption was stimulating in the knowledge of the flux and force representation of the Onsager principle in so far that with their aid the existence of a thermodynamic extremum principle of a form similar to the Gauss principle of mechanics could be examined. Since the Gauss principle is similar to the principle of least squares, i.e. it consists of the total squares of the difference of two quantities, it was plausible that with the dissipation potentials

$$\Phi \equiv \frac{1}{2} \sum_{i=1}^{f} R_{ii} J_i^2 \tag{4.38a}$$

and

$$\Psi \equiv \frac{1}{2} \sum_{i=1}^{f} R_{ii}^{-1} X_i^2 \tag{4.38b}$$

determined by the diagonal forms of the linear relations (4.6), the representations (4.26) and (4.27) of the Onsager principle could be given explicitly with the hope to obtain total squares. It is easy to demonstrate that the partial flux or force representations separately do not satisfy the expectations [51]. In other words, the expressions (4.26) and (4.27) created with the dissipation potentials (4.38) do not display the characteristic feature of the principle of least squares, which is typical of the Gauss principle.

The situation is different in the case of the universal form (4.28) whose development was described. Indeed, the universal form (4.28) of the principle of least dissipation of energy can be written with the dissipation potentials (4.38) as

$$\delta\left[\sum_{i=1}^{f} J_i X_i - \frac{1}{2} \sum_{i=1}^{f}\left(\frac{X_i^2}{R_{ii}} + R_{ii} J_i^2\right)\right] = \delta\left[-\frac{1}{2} \sum_{i=1}^{f} R_{ii}\left(J_i - \frac{X_i}{R_{ii}}\right)^2\right] = 0. \tag{4.39}$$

From this relation it follows—by also considering the negative sign—that in irreversible processes taking place in a continuum and being described by the linear Onsager theory, the quantity

$$C = \frac{1}{2} \sum_{i=1}^{f} R_{ii}\left(J_i - \frac{X_i}{R_{ii}}\right)^2 = \min. \tag{4.40}$$

It can be seen that this extremum principle, just as the Gauss principle of the least constraint, is already analogous to the principle of the least squares [43, 44, 58]. The quantity C defined in (4.40) can be interpreted similarly to the corresponding quantity of the Gauss principle as a "constraint", more precise as a "local constraint". Otherwise it is shown by the comparison of (4.40) to the Gauss principle that in thermodynamics the rôle of (inertial) masses is played by the resistances. The total "dictionary" of the corresponding mechanical and thermodynamical quantities is represented by the following analogies:

$$m \text{ mass} \leftrightarrow R \text{ resistance}$$

$$\boldsymbol{a} \text{ acceleration} \leftrightarrow \boldsymbol{J} \text{ velocity (flux)}$$

$$\boldsymbol{F} \text{ Newtonian force} \leftrightarrow \boldsymbol{X} \text{ thermodynamic force}.$$

Though in formal respect the analogy of the Gauss principle and the extremum principle (4.40) expressed by this "dictionary" is total, further conclusions on the essence of the relations between the fundamental principles of mechanics and thermodynamics cannot be drawn from it. It may be rather important that the above "dictionary" reflects one of the essential differences between the fundamental equations of thermodynamics from the basic equations of mechanics, and this is the one according to which in thermodynamics newtonian forces proportional to the mass and acceleration do not occur, but only general (not newtonian) forces, which are according to the linear kinematical constitutive equations (4.6) proportional to the resistances and the general velocities (fluxes).

It is proved by the foregoing that the principle analogous to the Gaussian differential principle of mechanics is valid also in thermodynamics and that this is a particular form of the general local principle (4.28). The minimum principle (4.40) can be applied to local thermodynamic constraint problems exactly in the same manner as the Gauss principle of the least constraint for the case of constraint problems in mechanics. The applications presented in the following will prove this in a striking form. Of course, with the application of (4.40) for actual cases, the linear kinematical equations will be used instead of the diagonal forms in the general forms (4.2) and (4.6). With the dissipation potentials (4.9) and (4.10) in this general case, we get form (4.28):

$$C \equiv \frac{1}{2} \sum_{i,k=1}^{f} R_{ik} \left(J_i - \sum_{s=1}^{f} L_{is} X_s \right) \left(J_k - \sum_{r=1}^{f} L_{kr} X_r \right) = \min. \quad (4.41)$$

This form can be practically applied on a wide scale already. Of course, in the lack of local constraints the minimum of the "constraint" C is zero, just as in the case of the Gauss principle. This is plausible, since the "constraint" C as it is seen from the comparison of (4.28) with (4.41) is the negative of the local OM function defined in (4.33), i.e. $C \equiv -\varphi$.

In this respect the maximum principle (4.35) and the minimum principle (4.41) are the alternative forms of the most general local differential principle of the least dissipation of energy. Anyway, in the case of the existence of the local constraints the minimum (4.41) is not zero, and thus the Gaussian principle of thermodynamics can be formulated in the following way (GYARMATI [50, 51, 52]):

In any thermodynamic system in the case of given thermodynamical free forces and in the case of given local constraint conditions, the only irreversible processes which take place are such that the "constraint" C is minimum for them.

Now let us give some practical applications of the principle which has been discussed by VERHÁS originally [59, 60, 61]. The following treatment is in some cases more general than the original one.

4. Applications of the Local Principle for Constraint Problems

The local extremum principle (4.41) allows the exact treatment of local thermodynamic constraint problems (particularly of diffusional, electrochemical, etc. ones) which are also important from practical respect. First of all let us examine a simple electrochemical problem where the actual form of Onsager's linear constitutive equations are determined with the help of the extremum principle (4.41).

Let us consider an electrolyte solution consisting of K components and let us assume that the electrical current density vanishes in each point of the system. Such cases are often encountered in diffusion and thermodiffusion electrochemical systems, when the diffusion forces are given by (3.72). The vanishing of the conductive current density (2.51), i.e.

$$\boldsymbol{i} = \sum_{k=1}^{K} e_k \boldsymbol{J}_k = 0 \tag{4.42}$$

means a local constraint, which should be taken into account with the application of the extremum principle (4.41). Hence, the determination of the linear kinematical equations valid in the case of our electrolyte system is an extremum problem owing to the restriction (4.42), which can be solved by using the minimum principle (4.41). Briefly, such a minimum of the C "constraint" of (4.41) shall be determined, which is compatible due to the vanishing of the conductive current density with the restriction (4.42) valid for the diffusion current densities $\boldsymbol{J}_k$.

The above extremum problem can immediately be solved by the Lagrangian multiplicator method. Multiplying (4.42) by a vector multiplicator $\boldsymbol{\lambda}$ and adding the result to the "constraint" C in (4.41), the partial derivations of the obtained expression with respect to the components $J_{k\alpha}$, $(\alpha = x_1, x_2, x_3)$ must vanish:

$$\frac{\partial}{\partial J_{k\alpha}}\left(C + \boldsymbol{\lambda} \cdot \sum_{k=1}^{K} e_k \boldsymbol{J}_k\right) = 0, \qquad (\alpha = x_1, x_2, x_3). \tag{4.43}$$

This condition can be rewritten by using (4.5) and (4.41) as

$$\sum_{i=1}^{K} (R_{ki} \boldsymbol{J}_i - \boldsymbol{X}_k + \boldsymbol{\lambda} e_k) = 0, \qquad (k = 1, 2, \ldots, K) \tag{4.44}$$

from which

$$\boldsymbol{J}_i = \sum_{k=1}^{K} L_{ik}(-\boldsymbol{\lambda} e_k + \boldsymbol{X}_k), \qquad (i = 1, 2, \ldots, K) \tag{4.45}$$

is obtained. Identifying the multiplicator $-\boldsymbol{\lambda}$ with the electric field strength $\boldsymbol{E} \equiv -\boldsymbol{\lambda}$, from (4.45) we get

$$\boldsymbol{J}_i = \sum_{k=1}^{K} L_{ik}(\boldsymbol{E} e_k + \boldsymbol{X}_k), \qquad (i = 1, 2, \ldots, K) \tag{4.46}$$

which is the linear relation wellknown in the case of electrolyte systems, if $\boldsymbol{X}_k$ is regarded the diffusion force (3.72). It is evident from this procedure that the electric force quantity $\boldsymbol{E} e_k$ occurring in (4.46) due to the "constraint" (4.42) besides the free force $\boldsymbol{X}_k$ may be interpreted as a *constraint force*. Therefore, in (4.46) the force

$$\boldsymbol{X}_k^0 = \boldsymbol{E} e_k + \boldsymbol{X}_k, \qquad (k = 1, 2, \ldots, K) \tag{4.47}$$

is the resultant of the thermodynamic free and constraint forces. It is perhaps needless to say that in electrolyte solutions the condition of zero current (4.42) can easily be realized and, e.g. the electrical potential produced can be measured by compensation. In these cases an external compensating potential difference must be applied to the electrolyte system, so that (4.42) be fulfilled in the course of the measurement, where the existence of the constraint force

$$\boldsymbol{X}_k' = \boldsymbol{E} e_k, \qquad (k = 1, 2, \ldots, K) \tag{4.48}$$

is locally ensured.

For the treatment of local constraint problems more general than the preceeding one, let us consider the most general set of the linear kinematical constitutive equations (4.2), given with scalar fluxes and forces. Let us assume that apart from the free forces X_k acting in the system also a local constraint exists, which yields a linear dependence

$$\sum_{k=1}^{f} a_k J_k = b \tag{4.49}$$

between the fluxes, where the coefficients a_k and b may still be functions of the local (equilibrium) state parameters.

To solve this general problem we can proceed in a manner analogous to the above. Let us denote the Lagrangian multiplicator by λ which is now a scalar. Using the minimum principle (4.41) the result

$$J_i = \sum_{k=1}^{f} L_{ik}(-\lambda a_k + X_k), \qquad (i = 1, 2, \ldots, f) \tag{4.50}$$

similar to (4.45) is obtained, where J_k and X_k denote the scalar components of the fluxes and free forces. The equation

$$\sum_{i,k=1}^{f} a_i L_{ik}(-\lambda a_k + X_k) = b \tag{4.51}$$

following from (4.50) and (4.49) serves for the determination of the Lagrangian multiplicator. This condition can be written with the introduction of the new quantities

$$a_k^0 = \sum_{i=1}^{f} a_i L_{ik} \tag{4.52a}$$

and

$$X_k^0 = -\lambda a_k + X_k \tag{4.52b}$$

in a concise form

$$\sum_{k=1}^{f} a_k^0 X_k^0 = b \tag{4.53}$$

similar to (4.49). The new forces X_k^0, similarly to the particular case (4.47), are the resultants of the constraint forces

$$X_k' = -\lambda a_k, \qquad (k = 1, 2, \ldots, f) \tag{4.54}$$

and the free forces X_k. Of course, the constraint forces are unknown as long as the Lagrangian multiplicator λ has not been determined. However, the determination of λ is always possible from (4.51), since according to the expression

$$\lambda = \frac{\sum_{i,k=1}^{f} a_i L_{ik} X_k - b}{\sum_{i,k=1}^{f} L_{ik} a_i a_k} = \frac{\sum_{k=1}^{f} a_k^0 X_k - b}{\sum_{k=1}^{f} a_k^0 a_k} \tag{4.55}$$

following from it, λ is determined by the minimum principle (4.41) and the constraint (4.49) in the case of given free forces X_k.

By taking relations (4.49) and (4.53) into consideration our results can be summarized as follows: *If in a system due to the existence of local constraints there is a linear dependence between the fluxes, then a similar dependence consists between the resultant (free + constraint) forces* X_k^0 *as well.*

This consequence can also be interpreted in an alternative way. In order to develop the alternative interpretation, the expression (4.55) of the multiplicator λ is substituted into the linear relation (4.50), whereupon the following expression is obtained

$$J_i = \sum_{k=1}^{f} L'_{ik} X_k - b \frac{\sum_{j=1}^{f} L_{ij} a_j}{\sum_{r,s=1}^{f} L_{rs} a_r a_s}, \qquad (i = 1, 2, \ldots, f) \qquad (4.56)$$

where the new

$$L'_{ik} = L_{ik} - \frac{\sum_{j,s=1}^{f} L_{ij} L_{ks} a_j a_s}{\sum_{r,s=1}^{f} L_{rs} a_r a_s}, \qquad (i, k = 1, 2, \ldots, f) \qquad (4.57)$$

conductivity coefficients have been introduced. According to (4.56), if $b = 0$, i.e. the linear dependence (4.49) is homogeneous, the fluxes J_k are determined by the conductivity coefficients L'_{ik} in terms of the free forces X_k. Now it is, of course, expected that the set of the coefficients L'_{ik} is not independent. Indeed, multiplying the coefficients L'_{ik} in (4.57) by a_i and summing, we get

$$\sum_{k=1}^{f} a_i L'_{ik} = 0. \qquad (4.58)$$

This relation is, just as (4.53), the consequence of the conditions (4.49) valid for the fluxes. Therefore, our results can be summarized as follows: *If a local constraint of the type* (4.49) *exists for the fluxes, its influence can be given either by the relation* (4.53) *valid for the forces* X_k *or by the restriction* (4.58) *valid for the transformed coefficients* L'_{ik}. Hence, the restrictions (4.53) and (4.58) are equivalent to one another and both are the result of the linear dependence (4.49).

If, instead of the inhomogeneous relation (4.49) we restrict ourselves to the homogeneous linear dependence which occurrs in the majority of cases, we shall have $b = 0$ in every given formula. The case of diffusion systems comprising of K components is an important example for such a homogeneous linear dependence, since for the diffusion current densities $\mathbf{J}_k$ the local constraint (1.43) is always valid. For such cases the following important theorem can be proved:

The validity of Onsager's reciprocal relations is not influenced by a linear homogeneous dependence valid amongst the fluxes.

As regards the proof of the theorem let us refer to the literature [3, 4]. However, it should be noted that there are also constraint problems, in which a local constraint for the forces instead of the fluxes is valid. Such a constraint is encountered in the case of mechanical equilibrium (2.118) fulfilled approximately in multi-component hydrodynamic systems. Indeed, as it was originally demonstrated by Prigo-

GINE [17], it follows from the Gibbs-Duhem relation (3.29) valid for isothermal cases

$$\sum_{k=1}^{K} \varrho_k (\nabla \mu_k)_T = \nabla p, \tag{4.59}$$

and from (2.118) that

$$\sum_{k=1}^{K} \varrho_k \{(\nabla \mu_k)_T - \boldsymbol{F}_k\} \equiv - \sum_{k=1}^{K} \varrho_k \boldsymbol{X}_k'' = 0, \tag{4.60}$$

where $\boldsymbol{X}_k''$ are the actual diffusion forces. Equation (4.60) represents a linear homogeneous local constraint amongst the diffusional (free) forces. In this case the Gaussian type minimum principle (4.41) can be applied just as in the cases of the constraint problems (4.42) or (4.49). Since this application is very simple, we will not discuss it here. Similarly, the following important theorem will be mentioned without proof [3].

If homogeneous linear relatonships exist between the fluxes as well as between the forces, the phenomenological coefficients are not uniquely defined, and Onsager's reciprocal relations are not necessarily fulfilled. However, it can, be proved that the coefficients can always be chosen in such a manner that the reciprocal relations shall remain valid.

The proof of this theorem is important from the viewpoint of applications and has been given by DE GROOT and MAZUR [3]. It should be noted that the free forces must be understood under the forces mentioned in the theorem. Hence, the local constraint

$$\sum_{k=1}^{f} b_k X_k = 0 \tag{4.61}$$

prescribed, in general, in the theorem refers to the free forces X_k, such as is represented particularly by (4.60). Care must be taken not to confuse the local constraint of the type (4.61) following a priori from the physical nature of the free forces with the relation type (4.53) valid for the resultant of the free + constraint forces (not even in the case of $b = 0$). The latter is a consequence of the condition (4.49) a priori stipulated for the fluxes by the physical circumstances.

5. The Global Forms of the Principle

Up to now we have treated the local forms of the variational principle of the least dissipation of energy, which are really differential principles. This is particularly evident from the Gaussian form, since the Gaussian principle of the least constraint can be considered a prototype of the differential principles [44, 58]. It is obvious to extend the validity of the local principle to a global one valid for a whole of a continuum, and this has been proposed by ONSAGER for the case of adiabatically isolated non-continuous systems and anisotropic heat conduction in flux representation [22, 46]. The general formulation of the global (or integrated) forms, both in flux and force representation, is the result of recent times (GYARMATI [50, 51]). In the following the global correspondent of all the representations of the local principles will be given.

The variation in the course of time of the total entropy of a continuum is given by

$$\dot{S} = \int_V \varrho \dot{s} \,\mathrm{d}V \tag{4.62}$$

following from (3.13). This entropy variation is composed according to (3.12) and (3.14) from the entropy quantity

$$\dot{S}^*[(\boldsymbol{J}_\mathrm{s})_{\boldsymbol{n}}] \equiv -\frac{\mathrm{d_r}S}{\mathrm{d}t} = \int_V \nabla\cdot \boldsymbol{J}_\mathrm{s}\,\mathrm{d}V = \oint_\Omega \boldsymbol{J}_\mathrm{s}\,\mathrm{d}\boldsymbol{\Omega} \tag{4.63}$$

exchanged along the boundary surface limiting the continuum from its surroundings (where $\dot{S}^*$ is a linear functional of $(\boldsymbol{J}_\mathrm{s})_{\boldsymbol{n}}$), further from the

$$\mathscr{P} \equiv \frac{\mathrm{d_i}S}{\mathrm{d}t} = \int_V \sigma\,\mathrm{d}V \geqq 0 \tag{4.64}$$

definite positive total entropy production. Introducing the global quantities of the local dissipation functions (4.9) and (4.10)

$$\boldsymbol{\Phi}(\boldsymbol{J}, \boldsymbol{J}) = \int_V \Phi(\boldsymbol{J}, \boldsymbol{J})\,\mathrm{d}V \geqq 0 \tag{4.65a}$$

and

$$\boldsymbol{\Psi}(\boldsymbol{X}, \boldsymbol{X}) = \int_V \Psi(\boldsymbol{X}, \boldsymbol{X})\,\mathrm{d}V \geqq 0, \tag{4.65b}$$

then with the help of these and of (4.64) the global OM function can be defined as

$$\mathcal{O} = \int_V \mathcal{o}\,\mathrm{d}V = \mathscr{P} - (\boldsymbol{\Psi} + \boldsymbol{\Phi}) = \dot{S} + \dot{S}^* - (\boldsymbol{\Psi} + \boldsymbol{\Phi}), \tag{4.66}$$

where it was also taken into account that the entropy balance (3.16) can be given with the aid of (4.62), (4.63) and (4.64) in the global form:

$$\dot{S} + \dot{S}^* = \mathscr{P} \geqq 0. \tag{4.67}$$

This form is, of course, nothing else than the relation (3.12) expressing the Carnot-Clausius theorem.

The global quantities allow to give the local extremum conditions in global forms as well. These conditions are, in different representations the following.

Flux representation:

$$\delta \int_V [\varrho \dot{s} + \nabla \cdot \boldsymbol{J}_s - \Phi]_{\boldsymbol{X}} \mathrm{d}V = \delta \int_V [\sigma - \Phi]_{\boldsymbol{X}} \mathrm{d}V = 0, \delta \boldsymbol{X} = 0, \delta \boldsymbol{J} \neq 0, \tag{4.68}$$

which can be given by using the relations (4.62) to (4.65a) also as

$$\delta[\dot{S} + \dot{S}^* - \boldsymbol{\Phi}]_{\boldsymbol{X}} = \delta[\mathscr{P} - \boldsymbol{\Phi}]_{\boldsymbol{X}} = 0, \delta \boldsymbol{X} = 0, \delta \boldsymbol{J} \neq 0. \tag{4.69}$$

This formulation is due to Onsager for the case of anisotropic heat conduction [22].

Force representation:

$$\delta \int_V [\varrho \dot{s} + \nabla \cdot \boldsymbol{J}_s - \Psi]_{\boldsymbol{J}} \mathrm{d}V = \delta \int_V [\sigma - \Psi]_{\boldsymbol{J}} \mathrm{d}V = 0, \delta \boldsymbol{J} = 0, \delta \boldsymbol{X} \neq 0, \tag{4.70}$$

which can be given by using the relations (4.62) to (4.65b) also as

$$\delta[\dot{S} + \dot{S}^* - \boldsymbol{\Psi}]_{\boldsymbol{J}} = \delta[\mathscr{P} - \boldsymbol{\Psi}]_{\boldsymbol{J}} = 0, \quad \delta \boldsymbol{J} = 0, \quad \delta \boldsymbol{X} \neq 0. \tag{4.71}$$

This representation is introduced first by GYARMATI [31, 50, 51].

The global form of the universal principle:

$$\delta \int_V [\varrho \dot{s} + \nabla \cdot \boldsymbol{J}_s - (\Psi + \Phi)] \mathrm{d}V = \delta \int_V [\sigma - (\Psi - \Phi)] \mathrm{d}V$$
$$= \delta \int_V \mathscr{o} \, \mathrm{d}V = 0, \quad \delta \boldsymbol{J} \neq 0, \quad \delta \boldsymbol{X} \neq 0, \tag{4.72}$$

which can alternatively be rewritten as

$$\delta[\dot{S} + \dot{S}^* - (\boldsymbol{\Psi} + \boldsymbol{\Phi})] = \delta[\mathscr{P} - (\boldsymbol{\Psi} + \boldsymbol{\Phi})] = \delta \mathscr{O} = 0,$$
$$\delta \boldsymbol{J} \neq 0, \quad \delta \boldsymbol{X} \neq 0, \tag{4.73}$$

on the basis of the relations (4.62) to (4.66).

In order to avoid superfluous repeating of the formulae we do not give the global principles following from the local formulations (4.20), (4.25), (4.30) and belonging to the above global extremum conditions. However, it should be pointed out that the requirement of the global conditions (4.68) to (4.73) for the case of the prescribed variations always ensures the existence of adequate maxima. This is an evident consequence of the fact that the total entropy production (4.64) and also the global dissipation potentials are according to the Carnot-Clausius theorem always positive definite quantities.

a) The Special Forms of the Principle for Adiabatically Isolated Systems. If a continuum of the volume V is adiabatically isolated along the boundary surface Ω, the hitherto general forms of the variational principle are reduced to special ones. According to the condition of the

Now let us give the original forms of the principle according to Onsager. If we describe the representations (4.76a) and (4.76b) by using (4.82) and (4.83), it can be seen that similarly to the local cases proved in detail, the linear relations (4.84) together with the reciprocal relations (4.85) are included in the corresponding variational principle. Thus, for instance, in the case of the flux representation originally described by ONSAGER [22] the principle

$$[\dot{S}(\alpha, \dot{\alpha}) - \Phi(\dot{\alpha}, \dot{\alpha})] = \max \tag{4.86}$$

is obtained. Even in this particular case the force representation (4.76b) was not given by ONSAGER. However, he and MACHLUP have recognized a particular form of the universal principle in connection with the fluctuation theory. We shall briefly outline the way leading to it.

In the case of simultaneous variation of fluxes and forces the universal principle (4.77) contains both groups of the linear constitutive equations (4.84) together with the total set of the reciprocal relations (4.85). Since by taking into account (4.81), (4.82) and (4.83), equation (4.77) can be given in a detailed form

$$\mathcal{O}^{\mathrm{ad}}(\alpha, \dot{\alpha}) = \{\dot{S}(\alpha, \dot{\alpha}) - [\Psi(X, X) + \Phi(\dot{\alpha}, \dot{\alpha})]\} = \max \tag{4.87}$$

(indicating also the independent variables explicitly) thus it is evident that the linear constitutive equations (4.84) are exactly the Euler-Lagrange equations of this variational principle. These equations are deterministic, as can easily be seen from the first order ordinary differential equations

$$\sum_{k=1}^{f} (\overline{R}_{ik}\dot{\alpha}_k + S_{ik}\alpha_k) = 0, \qquad (i = 1, 2, \ldots, f) \tag{4.88}$$

obtained as a combination of (4.82) and (4.84b), derived by ONSAGER and MACHLUP [46]. However, the deterministic character of the linear kinematical equations was destroyed by ONSAGER and MACHLUP in accordance with the thinking of the fluctuation theory. Namely, the original basic idea of Onsager was that the average decay of fluctuations in an adiabatically isolated "aged" system follow the ordinary macroscopic laws, for example Fourier's law, Fick's law, Ohm's law, etc. In this concept the (4.84) linear kinematical constitutive equations can be considered as mean laws, valid only for a certain period of time, which do not give an account of the fluctuations. Therefore the equations (4.84b) have been completed by ONSAGER and MACHLUP by the addition of the random force ε_i, i.e. instead of (4.84b) the Langevin type

$$\sum_{k=1}^{f} \overline{R}_{ik}\dot{\alpha}_k = X_i + \varepsilon_i, \qquad (i = 1, 2, \ldots, f) \tag{4.89}$$

stochastic equation was written [46]. Now the task was to decide, now the evolution of the parameters α_i in time can be determined under the influence of the random force ε_i. In a way, which cannot be specified here, with the use of a general theory of the Gaussian type Markoff processes, ONSAGER and MACHLUP obtained the result (see also [4, 47]) that the most probable path in time of the parameters α_i is determined by the form

$$\int_{t_1}^{t_2} \mathcal{O}^{\text{ad}}[\alpha(t), \dot{\alpha}(t)]\, \mathrm{d}t = \max \tag{4.90}$$

of the principle of the least dissipation of energy. It should be noted that ONSAGER and MACHLUP considered the principle (4.90) in spite of its universal form, only as one in which the variation must be exclusively taken with respect to the fluxes. On the other hand, it is also true that the formulation (4.90) depends, (in addition to the particular condition of the adiabatically isolated system-model) also on several particular requirements of the theory of stochastic processes (Gaussian distribution, Markovian character, stationarity, reversibility, etc.).

It is already evident from the above-mentioned objections that the productivity of the variational principle of the least dissipation of energy relies, on the one hand, upon the general local formulation of the principle, whereas, on the other hand, on the simultaneous variation with respect to the fluxes and forces. It will be seen later on that the force representation, i.e. the variation with respect to forces—always left out of consideration by Onsager and followers—which allows wide-range application of the principle. The advantage of the local forms, as compared to the global formulations, consists in the fact that the latter can be deduced from the former in the case of any system-model, but conversely this is not true.

b) The Special Forms of the Principle for Stationary Systems. The stationary state of open systems with respect to their surroundings (heat transfer, component transport, etc.) is characterized by the fact that the state parameters of such systems are constant in time. Therefore, in the stationary state of open systems

$$\dot{S} = 0, \tag{4.91}$$

i.e. the total entropy of the system is constant. On the other hand, from the balance (4.67) with the stationary condition (4.91) the reduced entropy balance

$$\dot{S}^*[(\boldsymbol{J}_s)_n] = \mathcal{P} \tag{4.92}$$

is obtained. According to this equation the stationary state can be maintained only if the amount of entropy flowing through the boundary surface of the system is replaced constantly by the entropy production and is according to (4.63) locally determined by the adequate stationary value

$$(\boldsymbol{J}_s)_{\boldsymbol{n}} = (\boldsymbol{J}_s)_{\boldsymbol{n}}^{\text{stac}} \tag{4.93}$$

of the normal component of the entropy current density.

By taking the foregoing into consideration and with the stationarity condition (4.91), the global forms (4.69), (4.71) and (4.73) of the variational principle are reduced to the special extremum principles

$$[\dot{S}^* - \boldsymbol{\Phi}]_{\boldsymbol{X}} = \max \tag{4.94a}$$

$$[\dot{S}^* - \boldsymbol{\Psi}]_{\boldsymbol{J}} = \max \tag{4.94b}$$

and

$$\mathcal{O}^{\text{stac}} = [\dot{S}^* - (\boldsymbol{\Psi} + \boldsymbol{\Phi})] = \max \tag{4.95}$$

where in the case of the universal form the special OM function $\mathcal{O}^{\text{stac}}$ valid for the stationary case has been introduced.

The stationary forms of the principle of least dissipation of energy can be transformed into somewhat more transparent forms. In case of flux representation, the transformation has been carried out first by Onsager [22]. Considering the local (4.9) and (4.10), further the global expressions (4.64) and (4.65), the following form of the entropy balance (4.92), valid for the stationary case, is obtained:

$$\dot{S}^* = \mathcal{P} = 2\,\boldsymbol{\Phi} = 2\boldsymbol{\Psi}, \tag{4.96}$$

with the help of which equations

$$2[\dot{S}^* - \boldsymbol{\Phi}] = \dot{S}^* \tag{4.97a}$$

$$2[\dot{S}^* - \boldsymbol{\Psi}] = \dot{S}^* \tag{4.97b}$$

are obtained. Since the left-hand side of these equations is—disregarding the factor 2—identical to the corresponding expression of the maximum principles (4.94) valid in stationary case, therefore, by fixation of the normal component of the entropy current density at the boundaries of the system according to (4.93), the value of $\dot{S}^*$ is also fixed, and thus from (4.94a) and (4.94b) the minimum principles

$$\boldsymbol{\Phi}(\boldsymbol{J}, \boldsymbol{J}) = \min, \quad \delta\boldsymbol{\Phi} = 0, \;\; \delta\boldsymbol{J} \neq 0 \tag{4.98}$$

and

$$\boldsymbol{\Psi}(\boldsymbol{X}, \boldsymbol{X}) = \min, \;\; \delta\boldsymbol{\Psi} = 0, \;\; \delta\boldsymbol{X} \neq 0 \tag{4.99}$$

are obtained, where the variational conditions have been written as well.

In the first case by following Onsager it can be said [22], that in the case of the stationarity conditions (4.91) and (4.93), the stationary distribution of the fluxes of irreversible processes is determined by the minimum principle (4.98), when the entropy current density is fixed on the boundary surface. It should be noted that in the case of hydro-thermodynamic systems in which heat flow $\boldsymbol{J}_q$ and flow of matter $\boldsymbol{J}_k$, $(k = 1, 2, \ldots, K)$ takes place through the boundary surface of the system, according to (3.67) in order to fix $(\boldsymbol{J}_s)_{\boldsymbol{n}}$, $(\boldsymbol{J}_q)_{\boldsymbol{n}}$ and every $(\boldsymbol{J}_k)_{\boldsymbol{n}}$, furthermore the temperature T and all the chemical potentials μ_k must be fixed. Similarly the minimum principle (4.99), formulated in force representation, determines the stationary distribution of the forces under the stationarity conditions (4.91) and (4.93). Of course, beyond the direct determination of the stationary distribution of the fluxes by the minimum principle (4.98) the stationary distribution of forces is also given by (4.98)—though in an indirect way—owing to the linear kinematical equations. Similarly, though the minimum principle (4.99) determines only the stationary distribution of the forces in a direct way, it gives—owing to the linear constitutive equations—the stationary distribution of the fluxes too. Perhaps it is needless to say that the Onsager variational principle is called the principle of least dissipation of energy, because in stationary cases, the principle is expressed by the adequate minimum of the dissipation potentials. More precise the denomination of the principle of least dissipation of energy originates from the principle (4.98), formulated with the minimum of the function $\boldsymbol{\Phi}$ [22].

Finally the universal principle (4.95) will be mentioned. Let us define a part of the local OM function of (4.33), which is solely determined by the sum of the local dissipation potentials, i.e.

$$g(\boldsymbol{J}, \boldsymbol{X}) = \Psi(\boldsymbol{X}, \boldsymbol{X}) + \Phi(\boldsymbol{J}, \boldsymbol{J}). \tag{4.100}$$

This function can be considered as *universal local potential*. If now the universal principle (4.95) valid for the stationary state is considered in the case of fixed $\dot{S}^*$, it can be seen that (4.95) is reduced to the extremum principle

$$\mathcal{G} = \int_V g \, \mathrm{d}V = \boldsymbol{\Psi} + \boldsymbol{\Phi} = \min, \tag{4.101}$$

where $\mathcal{G}$ is the global equivalent of the universal local potential (4.100). Of course, in the universal minimum principle (4.101), similarly to all principles given in universal form, the variation is carried out with respect to the fluxes and forces simultaneously, but independently of each other. It is also evident that the universal stationary principle (4.101) is considered a unified form of the special formulations (4.98) and (4.99). Indeed, the principle (4.101) becomes identical to (4.98) in case of constant forces and in the case of the variation of the fluxes, whereas in the case of the variation of the forces and in the case of constant fluxes it becomes identical to the particular form (4.99).

Herewith all the possible representations of the variational principle of least dissipation of energy was given, for stationary systems as well. In the following chapter, the principle of minimum production of entropy valid for stationary systems will be described. However, it will be seen that this principle is not a new one and not independent from the principle of least dissipation of energy, but it is a reformulation of the universal principle (4.101) in the "language of entropy production".

V. The Principle of Minimum Production of Entropy

The principle of minimum production of entropy was first formulated independently from the principle of least dissipation of energy by PRIGOGINE for the case of discontinuous systems [45, 17], and later on it was generalized by DE GROOT [8]. The total formulation of the principle has been carried out by GLANSDORFF and PRIGOGINE [64], who gave the extension of the principle for dissipative processes taking place in continuous systems by analysing the differential properties of the entropy production, moreover, they also outlined the validity limits.

Following the course of historical evolution, the principle will first be discussed for the case of discontinuous systems. Thus it will be possible to extend the content of the principle gradually from the simple cases to the more complicated ones, and, at the same time, it will be possible to introduce the concept of stationarity order in a transparent form. The latter is very productive from practical points of view. After this, the principle is formulated in a general form, and its relation to the principle of least dissipation of energy is cleared. The latter is a result of recent research (GYARMATI [51, 55]). It is demonstrated that the principle of minimum production of entropy is not a new and independent principle, but an alternative formulation in the "language of the entropy production" of the form of Onsager principle valid for stationary states. The condition of stationarity is strictly determined for dissipative processes with the aid of the principle and the stability of the stationary states is examined.

1. Stationary States of Discontinuous Systems

Though the flux representation (4.98) of the principle of least dissipation of energy referring to stationary states was already known in

1931, PRIGOGINE searched for another function — apart from Φ — having the characteristic property of an extremum in stationary states. Since the thermostatic equilibrium state is characterized by the entropy maximum and with zero entropy production, the assumption was plausible that the stationary states of an open system for certain fluxes is determined by the entropy production compatible with adequate auxiliary conditions. In such a way, the principle of minimum production of entropy became a general criterion of stationarity, and it is certain that this principle was the most productive principle of non-equilibrium thermodynamics at least up to recent times. First of all we discuss the principle for the case of a discontinuous system-model consisting of two homogeneous subsystems.

Let us consider a system consisting of two homogeneous (equilibrium) subsystems I and II, while the total system is a non-equilibrium one (Fig. 6). In such cases inhomogeneities occur only at the boundaries

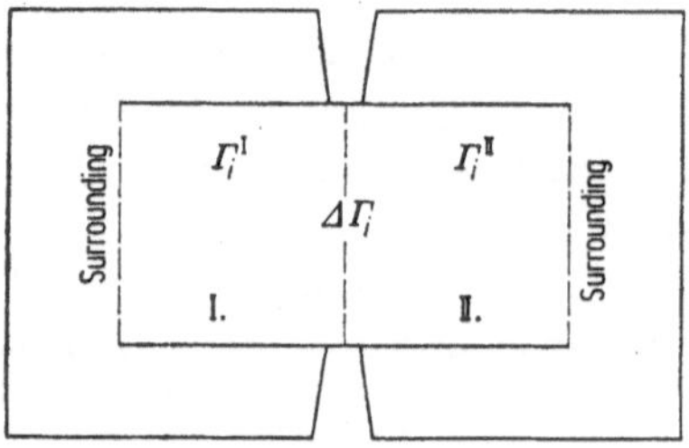

Fig. 6

of the subsystems, i.e. the thermodynamic forces $X_i \equiv \Delta\Gamma_i = \Gamma_i^{\mathrm{I}} - \Gamma_i^{\mathrm{II}}$ within the individual subsystems do not depend on the space coordinates, hence, they exist only along the boundary surfaces of the two subsystems. Let us also assume that the subsystems are open with respect to their sorroundings, in the sense that in certain stationary states a number j of the fluxes does not vanish between the subsystems and their surroundings—and, of course, also along the boundary surfaces of the subsystems—, but is of constant value in time, i.e.

$$J_i^0 = \text{const}, \qquad (i = 1, 2, \ldots, j). \tag{5.1}$$

In order to ensure the constancy of the fluxes in question it is necessary that between the subsystems and their surroundings the conjugated forces be kept at a constant value

$$X_i^0 = \text{const}, \qquad (i = 1, 2, \ldots, j) \tag{5.2}$$

with the aid of artificial constraints (thermostats, etc.). Of course, the constancy of all the state parameters of the total system consisting of two subsystems is ensured by the condition (5.1) and (5.2) only in the case, if $j = f$, i.e. for instance, if all the independent forces along the boundaries of the system are artificially fixed. If this is not the case, i.e. $j < f$, the necessary condition of the stationary state is determined with the aid of the principle of minimum production of entropy.

In case of choosing a discontinuous system as the model of analysis, the apparatus given in Chapter IV sect. 5 point a) can be applied for adiabatically isolated systems without any considerable change (see, for instance, [8, 38, 49]), and consequently by using (4.81) and (4.84a) the entropy production of the total system can be given in force representation as

$$\mathscr{P} = \sum_{i,k=1}^{f} \overline{L}_{ik} X_i X_k . \tag{5.3}$$

Let us consider its minimum under the conditions (5.2) assuming that from among all the forces $X_1, X_2, \ldots, X_f$ only the $X_1, X_2, \ldots, X_j$ are artificially fixed along the boundaries of the system, whereas the others $X_{j+1}, X_{j+2}, \ldots, X_f$ can change freely. Under the circumstances dealt with it is required by the necessary condition of the existence of a state of minimum entropy production that the condition

$$\frac{\partial \mathscr{P}}{\partial X_i} = 0, \qquad (i = j+1, j+2, \ldots, f) \tag{5.4}$$

be fulfilled for the forces $X_{j+1}, X_{j+2}, \ldots, X_f$, which are not artificially kept at constant values by means of external constraints along the boundary surfaces of the system. Since $\mathscr{P}$ is positive definite, the existence of such a minimum entropy production is ensured by the (5.4) extremum condition, which is compatible with the auxiliary conditions (5.2). The necessary condition (5.4) of the minimum in question leads with (5.3) to the result that

$$\frac{\partial \mathscr{P}}{\partial X_i} = \sum_{k=1}^{f} (\overline{L}_{ik} + \overline{L}_{ki})\, X_k = 2 \sum_{k=1}^{f} \overline{L}_{ik} X_k = 2J_i = 0,$$

$$(i = j+1, j+2, \ldots, f) \tag{5.5}$$

is valid, if the validity of the basic postulates of the linear Onsager theory, i.e. the validity

1. of the linear kinematical equations: $\left(J_i = \sum_{k=1}^{f} \overline{L}_{ik} X_k\right)$,

2. of the constancy of the coefficients: $(\overline{L}_{ik} = \text{const})$,

3. of the existence of the reciprocal relations: $(\overline{L}_{ik} = \overline{L}_{ki})$.

are assumed.

The stationary state of the system is unambigously ensured by the above conditions, since in the state of minimum entropy production compatible with the conditions (5.1) and (5.2), the constancy of the first fluxes and forces number j is ensured by keeping the forces $X_1, X_2, \ldots, X_j$ artificially at constant values. Whereas the constancy of the remaining forces $X_{j+1}, X_{j+2}, \ldots, X_f$ is ensured by the vanishing of the conjugated fluxes according to (5.5) and to the linear constitutive equations, thus the constancy of each parameter in time is ensured by the given conditions, i.e. the system is stationary.

The given conditions of stationarity and thus the recognition of the principle of minimum production of entropy has been proposed by PRIGOGINE for the case of one single fixed force [45, 17]. The general formulation is due to DE GROOT [8], and it was also he who introduced the productive concept of the order of stationarity, which relys upon (5.5), and by which a more transparent classification of the stationary systems is enabled. According to DE GROOT [8]:

A thermodynamic system is in a stationary state of the order j if from among f independent forces, j are artificially fixed, and at the same time the system is in a state of minimum entropy production. In this case, the fluxes conjugated to the artificially non fixed forces vanish, and thus each state parameter of the system takes a value constant in time.

Two particular cases must be differentiated. First, when all the forces are artificially fixed, i.e. $j = f$. This is the case which corresponds to a stationary state artificially forced in each detail; however, this case is uninteresting both from theoretical and practical points of view. In the second no force is fixed, i.e. $j = 0$, but the condition of the minimum entropy production is fulfilled. Such a system can only be a closed equilibrium system, because from condition (5.5) it follows that all the fluxes vanish and thus also the entropy production will be zero. Hence, the stationary state of zero order corresponds to the thermostatic equilibrium state of a closed system, where the entropy production is zero.

The necessary condition (5.4) of the states of minimum entropy production determines, together with the adequate auxiliary conditions, the stationary state of any order of the system, if the fundamental postulates of the linear Onsager theory were valid. Let us examine now in a very simple way the condition sufficient the existence of the minimum. This condition is related to the stability of the stationary states, by which the Le Chatelier-Braun principle, wellknown from thermostatics, can be extended to irreversible processes.

Let us consider a small perturbation δX_m of the non-fixed forces

$$X_m = X_m^0 + \delta X_m, \qquad (m = j + 1, j + 2, \ldots, f) \tag{5.6}$$

where X_m^0 is the stationary value. The linear kinematical constitutive equations referring to the conjugated fluxes

$$J_n = \sum_{m=1}^{f} \overline{L}_{nm} X_m = \sum_{m=1}^{f} \overline{L}_{nm} X_m^0 + \sum_{m=j+1}^{f} \overline{L}_{nm}\, \delta X_m = J_n^0 + \delta J_n$$

$$(n = j + 1, j + 2, \ldots, f) \tag{5.7}$$

are reduced to the form

$$J_n = \sum_{m=j+1}^{f} \overline{L}_{nm}\, \delta X_m = \delta J_n, \quad (n = j + 1, j + 2, \ldots, f) \tag{5.8}$$

since the non perturbated fluxes vanish, i.e.

$$J_n^0 = \sum_{m=1}^{f} \overline{L}_{nm} X_m^0 = 0, \qquad (n = j + 1, j + 2, \ldots, f) \tag{5.9}$$

is fulfilled because of the condition (5.5). By taking these into account, the entropy production can be split into two parts:

$$\mathcal{P} = \sum_{i,k=1}^{f} \overline{L}_{ik} X_i^0 X_k^0 + \sum_{n,m=j+1}^{f} \overline{L}_{nm}\, \delta X_n\, \delta X_m, \tag{5.10}$$

where

$$\mathcal{P}^0 = \sum_{i,k=1}^{f} \overline{L}_{ik} X_i^0 X_k^0 \geqq 0 \tag{5.11}$$

is the minimum entropy production corresponding to the stationary state, whereas

$$\delta\mathcal{P} = \sum_{n,m=j+1}^{f} \overline{L}_{nm}\, \delta X_n\, \delta X_m \geqq 0 \tag{5.12}$$

is a unique measure of the deviation from the stationary state of the entropy production. The minimum of $\mathcal{P}$ in stationary states is ensured by the positive definite property of $\delta\mathcal{P}$. Considering (5.8), expression (5.12) can be rewritten as

$$\delta\mathcal{P} = \sum_{n=j+1}^{f} \delta J_n\, \delta X_n \geqq 0, \tag{5.13}$$

the detailed analysis provides us with the possibility of the examination of stability of the stationary state and leads to the above-mentioned generalization of the Le Chatelier-Braun principle [8, 31].

When a closed thermodynamic system is allowed to pass into its equilibrium state, the entropy production decreases in time, and at the same time, the entropy S of the system approaches to the maximum value corresponding to the equilibrium state. If the system reaches equilibrium, the

$$\mathscr{P} = 0 \tag{5.14a}$$

$$S_0 = \max \tag{5.14b}$$

conditions determine the statical equilibrium state.

The situation is similar to an open system undergoing a change ending in the stationary state of the j-th order. In such cases the deviation $\delta\mathscr{P}$ from the $\mathscr{P}^0$ minimum value of entropy production determining the stationary state decreases in the course of time, whereas the conditions

$$\delta\mathscr{P} = 0 \tag{5.15a}$$

$$\mathscr{P}^0 = \min \tag{5.15b}$$

characterize the stationary state itself. In order to describe the approach of a system towards the stationary state, the differential properties of the entropy production must be analyzed in more detail.

The following important theorem—which includes several partial ones—was first proved by GLANSDORFF and PRIGOGINE in 1954 [64].

1) From the expression of the entropy production, which is bilinear in terms of the independent fluxes and forces

$$\mathscr{P} = \sum_{i=1}^{f} J_i X_i \geqq 0 \tag{5.16}$$

it follows that we can always decompose the time change of $\mathscr{P}$ into two parts

$$\mathrm{d}\mathscr{P} = \sum_{i=1}^{f} J_i \,\mathrm{d}X_i + \sum_{i=1}^{f} X_i \,\mathrm{d}J_i . \tag{5.17}$$

The first part is the consequence of the variation of forces, whereas the second one is related to the change of the fluxes, i.e.

$$\mathrm{d}\mathscr{P} = \mathrm{d}_X\mathscr{P} + \mathrm{d}_J\mathscr{P}, \tag{5.18}$$

where the usual denotations

$$\mathrm{d}_X\mathscr{P} \equiv \sum_{i=1}^{f} J_i \,\mathrm{d}X_i \tag{5.19a}$$

$$\mathrm{d}_J\mathscr{P} \equiv \sum_{i=1}^{f} X_i \,\mathrm{d}J_i \tag{5.19b}$$

have been introduced.

2) In the case of the validity of the basic postulates of the linear theory it can be proved that the variation of the entropy production due to the variation of the forces is equal to the variation of the entropy

production due to the variation of the fluxes, and the partial variations are separately equal to half of the total change of the entropy production, i.e.

$$2\,\mathrm{d}_X\mathscr{P} = 2\mathrm{d}_J\mathscr{P} = \mathrm{d}\mathscr{P}. \tag{5.20}$$

This theorem can be proved immediately. Namely, in the course of the transformations leading to the equality

$$\begin{aligned} \mathrm{d}_X\mathscr{P} &\equiv \sum_{i=1}^{f} J_i\,\mathrm{d}X_i \\ &= \sum_{i,k=1}^{f} \overline{L}_{ik}X_k\,\mathrm{d}X_i = \sum_{i,k=1}^{f} X_k(\overline{L}_{ik}\,\mathrm{d}X_i) = \sum_{k,i=1}^{f} X_k(\overline{L}_{ki}\,\mathrm{d}X_i) \\ &= \sum_{k=1}^{f} X_k\,\mathrm{d}J_k \equiv \mathrm{d}_J\mathscr{P} \end{aligned} \tag{5.21}$$

only the three basic postulates of the linear theory were used. However, it should be emphasized that in more general cases, in which from among the three basic postulates of the linear theory only one is not valid, the equality (5.21) does not hold. Thus neither the validity of (5.20) can be proved.

3) If (5.17) is written in the form

$$\frac{\mathrm{d}\mathscr{P}}{\mathrm{d}t} = \sum_{i=1}^{f} J_i\frac{\mathrm{d}X_i}{\mathrm{d}t} + \sum_{i=1}^{f} X_i\frac{\mathrm{d}J_i}{\mathrm{d}t} = \frac{\mathrm{d}_X\mathscr{P}}{\mathrm{d}t} + \frac{\mathrm{d}_J\mathscr{P}}{\mathrm{d}t} \tag{5.22}$$

it can be seen that, the evolution of the entropy production can be determined in each case, when the differential equations, governing the variation of the forces and fluxes in time, are known. The essence of the Prigogine and Glansdorff theorem is expressed by the fact that in the case of purely dissipative processes

$$\frac{\mathrm{d}\mathscr{P}}{\mathrm{d}t} \leqq 0 \tag{5.23}$$

is valid, i.e. the entropy production of the system decreases in time, if the system is approaching its stationary state stipulated by the time-independent boundary conditions. The equality holds in the stationary state, when the entropy production takes a minimum value compatible with the time-independent boundary conditions.

The condition (5.23) expresses the principle of minimum production of entropy in the case of different system-models in different ways and it can be proved with a different accuracy. A simple and total proof of (5.23) can be given for discontinuous systems, if the validity of the basic postulates of the linear theory is assumed. Since, in such a case (5.21) is valid, with its aid and by using (5.22) it can be shown that

$$\frac{\mathrm{d}\mathscr{P}}{\mathrm{d}t} = 2\frac{\mathrm{d}_X\mathscr{P}}{\mathrm{d}t} = 2\frac{\mathrm{d}_J\mathscr{P}}{\mathrm{d}t} \leqq 0 \tag{5.24}$$

is valid in every case, when an of

$$\frac{d_X \mathscr{P}}{dt} = \sum_{i=1}^{f} J_i \frac{dX_i}{dt} \leqq 0 \tag{5.25a}$$

or

$$\frac{d_J \mathscr{P}}{dt} = \sum_{i=1}^{f} X_i \frac{dJ_i}{dt} \leqq 0 \tag{5.25b}$$

(possibly independently from each other) can be proved. The confirmation of (5.25a) for the case of a multi-component, non-isothermal and discontinuous system is due to Ono [48], whereas a general confirmation is accounted to de Groot and Mazur [3]. In the case of discontinuous systems, the general and simultaneous confirmation of (5.25a) and (5.25b) was given with the direct application of the differential equations valid for the forces and fluxes by Gyarmati and Oláh [63].

2. Formulation of the Principle for Continua

With the above presentation of the principle of the minimum entropy production, a global formalism was used, which can be applied in a direct manner only for discontinuous systems. The principle, however, can easily be reformulated also in the case of continua, if the form

$$\mathscr{P} = \int_{V^0} \sigma \, dV^0 = \int_{V^0} \left(\sum_{i=1}^{f} J_i X_i \right) dV^0 \tag{5.26}$$

of the entropy production is used, where J_i and X_i are the local fluxes and forces. It should be emphasized that in (5.26) the integral must be taken over the volume elements dV^0 expressed in terms of Euler coordinates, because the time derivative of the total entropy production can be written only in this case in the form

$$\frac{d\mathscr{P}}{dt} = \int_{V^0} \frac{\partial_X \sigma}{\partial t} dV^0 + \int_{V^0} \frac{\partial_J \sigma}{\partial t} dV^0$$

$$= \int_{V^0} \left(\sum_{i=1}^{f} J_i \frac{\partial X_i}{\partial t} \right) dV^0 + \int_{V^0} \left(\sum_{i=1}^{f} X_i \frac{\partial J_i}{\partial t} \right) dV^0 \tag{5.27}$$

originally used by Glansdorff and Prigogine [64, 3].

If the basic postulates of the linear theory are valid, the relation similar to (5.21)

$$\frac{\partial_X \sigma}{\partial t} = \frac{\partial_J \sigma}{\partial t} \tag{5.28}$$

is also valid for the variation of the local entropy production, by the use of which (5.24) can again be written for the variation of the global production in time. However, two considerable differences can be seen between the application of the principle for discontinuous cases and for continua.

The first difference arises from the fact that in the case of continua it is only the relation

$$\frac{\mathrm{d}_X \mathscr{P}}{\mathrm{d}t} \equiv \int\limits_{V^0} \frac{\partial_X \sigma}{\partial t} \mathrm{d}V^0 = \int\limits_{V^0} \left(\sum_{i=1}^{f} J_i \frac{\partial X_i}{\partial t} \right) \mathrm{d}V^0 \leqq 0 \tag{5.29}$$

related to the variation of forces which can be confirmed directly by the differential equations (the equation of heat conduction, diffusion, etc.) governing irreversible transport processes. Therefore, if more general cases are considered, in which some of the basic postulationes of the linear theory are not valid (for example the coefficients are not constant, but they are the functions of the state parameters) and thus (5.28) is not fulfilled, nothing can be said about the entropy variation

$$\frac{\mathrm{d}_J \mathscr{P}}{\mathrm{d}t} \equiv \int\limits_{V^0} \frac{\partial_J \sigma}{\partial t} \mathrm{d}V^0 = \int\limits_{V^0} \left(\sum_{i=1}^{f} X_i \frac{\partial J_i}{\partial t} \right) \mathrm{d}V^0 \tag{5.30}$$

arising from the variation of fluxes. In such cases the validity of (5.24) is not guaranteed at all. Therefore the stationary state does not necessarily correspond to the state of some minimum entropy production. However, (5.29) can be proved also in more general cases, at least for purely dissipative processes. Briefly, the principle of minimum production of entropy (5.24) is strictly valid only within the frames of the linear theory, whereas (5.29) is valid in the whole realm of thermodynamics of irreversible processes, i.e. for the more general non-linear cases as well. The foregoing will be proved later by examples.

In the case of continua, another difficulty results from the particular structure of the differential equations governing the irreversible transport processes. Namely, the equations of heat conduction and diffusion are such partial differential equations, by which the stationary states can be directly characterized in the case of the vanishing of the local (partial) time derivatives. The validity of (5.29) can be confirmed by these equations for purely dissipative processes. However, a particular difficulty is encountered, if in the case of viscous flow also the convective mechanical motion should be taken into consideration. For this case, exactly as in the preceeding one leading to non-linear problems, the principle of minimum production of entropy cannot be applied without further considerations. Nevertheless, these difficulties recently urged Prigogine and Glansdorff to generalize the principle of minimum production of entropy and to develop a general evolution criterion of macroscopic physics with the help of the so-called local potential theory [65, 66, 67, 68, 69, 70].

3. The Relation between Onsager's and Prigogine's Principle

The relation between the principle of minimum production of entropy and the principle of least dissipation of energy was up to recent times quite unclarified. This circumstance was brought about by the non-total development of the Onsager principle and its inapplicability for the solution of practical problems, together with the fact that PRIGOGINE recognized his principle in a way independent of the former. According to Ono's already cited statement [48], the essential part of the Onsager principle is the variation with respect to the fluxes, whereas, the essential part of the Prigogine principle must be seen in the simultaneous variation with respect to the fluxes and forces. However, in the knowledge of the force representation of the Onsager principle the conclusion can be drawn that for the establishment of the relation between the two principles the key is included in this representation. This is even more evident, since in the expression (5.3), leading to the recognition of the principle of minimum production of entropy and serving as a basis for the first formulation of this principle, the force representation of the entropy production is applied. By this fact it is not only proved that the knowledge of the force representation is indispensable with the clarification of the relations in question, but the enhanced applicability of the force representation (5.29) as compared to (5.30) also suggests that force representations are, in general, more productive than flux representations.

Based upon the most recent investigations (GYARMATI [51], and particularly [55]) it will be demonstrated that, the principle of minimum production of entropy is not an independent principle of non-equilibrium thermodynamics, but rather is only an alternative reformulation of the Onsager principle valid for stationary cases.

Consider the force representation of the principle of least dissipation of energy valid for the stationary state (4.99). From the following form of the dissipation potential

$$\Psi \equiv \frac{1}{2} \sum_{i,k=1}^{f} \bar{L}_{ik} X_i X_k \geqq 0 \qquad (5.31)$$

(referring, for the sake of simplicity to discontinuous systems), furthermore, from (4.99) by using the auxiliary condition (5.2), it follows that in the stationary case

$$\frac{\partial \Psi}{\partial X_i} = \sum_{i=1}^{f} \bar{L}_{ik} X_k = J_i = 0, \quad (i = j+1, j+2, \ldots, f) \qquad (5.32)$$

is valid for each flux conjugated to the artificially non fixed forces. This expression is in each respect equivalent to (5.5) given in terms of the entropy production, and even more elegant than (5.5), since the superfluous factor 2 does not occur. Our result is so trivial and it is based only upon the existence of the force representation (4.99), so that repeating each theorem already formulated with the entropy production becomes superfluous.

Let us examine the problem how the theorems 1), 2), and 3) given by Glansdorff and Prigogine can be reformulated with the representations (4.98), (4.99) and (4.101) valid for the stationary case of the principle of least dissipation of energy. According to the spirit of the field theory (and using the variational symbol δ) in the case of simultaneous variation of the forces and fluxes, the local variation of the entropy production can be written as

$$\delta\sigma = \sum_{i=1}^{f} J_i\,\delta X_i + \sum_{i=1}^{f} X_i\,\delta J_i \equiv \delta_X\sigma + \delta_J\sigma. \tag{5.33}$$

Since, rigorously within the frame of the linear theory, the relations

$$\delta_X\sigma = \delta_J\sigma \tag{5.34a}$$

and

$$\delta\sigma = 2\delta_X\sigma = 2\delta_J\sigma \tag{5.34b}$$

are valid, the principle of minimum production of entropy can be expressed in the most general way by the extremum

$$\mathscr{P} = \int_{V^0} \sigma\,\mathrm{d}V^0 = \min, \tag{5.35}$$

or by the extremum condition

$$\delta\mathscr{P} = \int_{V^0} \sum_{i=1}^{f} (J_i\,\delta X_i + X_i\,\delta J_i)\,\mathrm{d}V^0 \leqq 0, \tag{5.36}$$

if the corresponding boundary conditions are time-independent. Here the equality holds for a stationary state compatible with the boundary conditions. The expressions (5.33) to (5.36) represent the most general formulation of the principle of minimum production of entropy.

Now let us consider the universal form (4.101) of the principle of least dissipation of energy referring to the stationary state. This minimum principle is in the case of the simultaneous variation of the forces and fluxes equivalent to the extremum condition

$$\begin{aligned}\delta\mathscr{G} &= \delta\frac{1}{2}\int_{V^0}\left[\sum_{i,k=1}^{f} L_{ik}X_iX_k + \sum_{i,k=1}^{f} R_{ik}J_iJ_k\right]\mathrm{d}V^0\\ &= \int_{V^0}\left[\sum_{i,k=1}^{f} L_{ik}X_k\,\delta X_i + \sum_{i,k=1}^{f} R_{ik}J_k\,\delta J_i\right]\mathrm{d}V^0 \leqq 0\end{aligned} \tag{5.37}$$

which leads with the linear kinematical constitutive equations (4.2) and (4.6) to the condition

$$\delta \mathscr{G} = \int_{V^0} \sum_{i=1}^{f} (J_i \, \delta X_i + X_i \, \delta J_i) \, \mathrm{d}V^0 \leqq 0 \tag{5.38}$$

where the equality holds, just as in (5.36) for stationary cases. Again it has been found that the principle of minimum production of entropy does not include any new and independent idea with respect to the principle of least dissipation of energy. In other words, the Prigogine principle is nothing else, than a reformulation of the universal form of the Onsager principle valid for the stationary case in another language. The basis of the "dictionary" is formed by the following mutual conjugation

$$g \leftrightarrow \sigma, \qquad \mathscr{G} \leftrightarrow \mathscr{P} \tag{5.39}$$

between the universal dissipation potential (4.100) and the entropy production, which is based upon the equivalence of the description of irreversible processes of the parameter systems $\{J_1, J_2, \ldots, J_f\}$ and $\{X_1, X_2, \ldots, X_f\}$ respectively. Hence, whenever with the description of the stationary states the principle of minimum production of entropy is used, we are speaking in the "language" of the $\mathscr{P}$ function. Whereas, if the function $\mathscr{G}$ is considered as fundamental there is the question of the formulation of the principle of least dissipation of energy valid for the stationary case. The "dictionary" between the two "languages" is given in local form by

$$\delta\sigma \leftrightarrow \delta g, \qquad \delta_X \sigma \leftrightarrow \delta \Psi, \qquad \delta_J \sigma \leftrightarrow \delta \Phi, \tag{5.40}$$

whereas by

$$\delta \mathscr{P} \leftrightarrow \delta \mathscr{G}, \qquad \delta_X \mathscr{P} \leftrightarrow \delta \boldsymbol{\Psi}, \qquad \delta_J \mathscr{P} \leftrightarrow \delta \boldsymbol{\Phi}, \tag{5.41}$$

in the case of global formulation.

4. Applications

Now let us demonstrate also by practical examples that the treatment of the stationary states, heretofore always employing the principle of minimum production of entropy, can also be treated with the adequate form of the Onsager principle. Only the stationary states of dissipative systems will be analysed for which the linear theory can be applied to a good approximation, i.e. the three basic postulates of the Onsager theory are valid.

a) Heat Conduction in Solid Bodies. Let us consider a one-component isotropic and solid body whose thermal expansion can be neglected, and along the boundary surfaces of which a time-independent temperature distribution is maintained. For such a body the internal energy balance must be given in the reduced form following from (3.36)

$$\varrho \frac{\partial u}{\partial t} = \varrho c_{\mathrm{v}} \frac{\partial T}{\partial t} = - \nabla \cdot \boldsymbol{J}_{\mathrm{q}}, \tag{5.42}$$

where c_{v} is the specific heat at constant volume. Fourier's linear law (3.96) can also be given in the alternative forms

$$\boldsymbol{J}_{\mathrm{q}} = -\lambda \nabla T = L_{\mathrm{qq}} \nabla \left(\frac{1}{T}\right), \quad L_{\mathrm{qq}} = T^2 \lambda. \tag{5.43}$$

The total dissipation potential of the system expressed with the latter linear law given in the entropy picture will be

$$\boldsymbol{\Psi} = \int_{V^0} \Psi \, \mathrm{d} V^0 = \int_{V^0} \frac{L_{\mathrm{qq}}}{2} \left(\nabla \frac{1}{T}\right)^2 \mathrm{d} V^0 \tag{5.44}$$

which is, of course, half of the total entropy production

$$\mathscr{P} = \int_{V^0} \sigma \, \mathrm{d} V^0 = \int_{V^0} (\boldsymbol{J}_{\mathrm{q}} \cdot \boldsymbol{X}_{\mathrm{q}}) \, \mathrm{d} V^0 = \int_{V^0} L_{\mathrm{qq}} \left(\nabla \frac{1}{T}\right)^2 \mathrm{d} V^0. \tag{5.45}$$

Now let us determine the temperature distribution which corresponds to a minimum of $\boldsymbol{\Psi}$ according to the stationary form (4.99) of the principle of least dissipation of energy. The solution of this problem is obtained by the variational task

$$\delta \boldsymbol{\Psi} = \delta \int_{V^0} \frac{L_{\mathrm{qq}}}{2} \left(\nabla \frac{1}{T}\right)^2 \mathrm{d} V^0 = 0, \tag{5.46}$$

where the variation with respect to the temperature δT must be taken at the boundary to be zero. This variational task leads to the Euler-Lagrange equation

$$\Delta \left(\frac{1}{T}\right) = 0 \tag{5.47}$$

which is the Laplace equation determining the stationary temperature distribution. Indeed, by considering the linear law (5.43), instead of (5.47)

$$\nabla \cdot \boldsymbol{J}_{\mathrm{q}} = 0 \tag{5.48}$$

can be written by which from (5.42)

$$\frac{\partial T}{\partial t} = 0 \tag{5.49}$$

follows, which is the direct condition of the stationary temperature distribution.

It is shown by the above example that the stationary temperature distribution is uniquely given by a minimum of the dissipation potential determined with the variational condition (5.46). Of course, the same is true for the minimum of the entropy production (5.45), and also (5.47) which follows from $\delta \mathscr{P} = 0$ (disregarding only the aesthetically disturbing factor 2). In the literature heretofore the stationary temperature distribution has always been determined from the principle of minimum production of entropy [64, 3]. The fact, however, is indisputable that in the case of the application of the expression of entropy production (5.45), though in an implicit way, there is a question of a direct application of the dissipation potential Ψ depending on the forces.

It should be noted that the above-mentioned problem can also be treated with the dissipation potential

$$\Psi^{**} \equiv \int_{V^0} \frac{\lambda}{2} (\nabla T)^2 \,\mathrm{d}V^0, \qquad \Psi^{**} \equiv T^2 \Psi \tag{5.50}$$

including directly the ordinary heat conductivity coefficient λ, (which will be called the dissipation potential given in the Fourier picture). From the corresponding variational condition $\delta \Psi^{**} = 0$ now the Laplace equation

$$\Delta T = 0 \tag{5.51}$$

—which refers directly to the temperature—is obtained. This equation is in every respect equivalent to (5.47).

Now it will be shown that the stationary state determined by the minimum principles is stable with respect to local perturbations of temperature. If we differentiate (5.44) with respect to time, we get

$$\frac{\mathrm{d}\Psi}{\mathrm{d}t} = \int_{V^0} L_{\mathrm{qq}} \nabla\left(\frac{1}{T}\right) \cdot \nabla\left(\frac{\partial}{\partial t}\frac{1}{T}\right) \mathrm{d}V^0 \tag{5.52}$$

which, after a transformation with (5.43) and after partial integration, becomes

$$\frac{\mathrm{d}\Psi}{\mathrm{d}t} = \int_{V^0} \boldsymbol{J}_{\mathrm{q}} \cdot \nabla\left(\frac{\partial}{\partial t}\frac{1}{T}\right) \mathrm{d}V^0 = \int_{\Omega^0} \left(\frac{\partial}{\partial t}\frac{1}{T}\right) \boldsymbol{J}_{\mathrm{q}} \cdot \mathrm{d}\boldsymbol{\Omega}^0 - \int_{V^0} \left(\frac{\partial}{\partial t}\frac{1}{T}\right) \nabla \cdot \boldsymbol{J}_{\mathrm{q}} \,\mathrm{d}V^0. \tag{5.53}$$

Here the surface integral vanishes if the temperature along the boundaries of the system is fixed. Taking this into account and with the use of the energy balance (5.42) we obtain the conclusion that

$$\frac{\mathrm{d}\Psi}{\mathrm{d}t} = -\int_{V^0} \varrho \frac{c_{\mathrm{v}}}{T^2} \left(\frac{\partial T}{\partial t}\right)^2 \mathrm{d}V^0 \leq 0, \tag{5.54}$$

since every quantity ϱ, T, c_{v} in the integrand is positive. This result means that during the evolution of the system towards the stationary state the dissipation potential is always diminishing in the course of time until the system does not attain its stationary state compatible with the boundary conditions. In other words, the stationary state characterizable with the minimum of Ψ is stable.

As a matter of course, the physical content of the inequality (5.54) may be translated into the "language of the entropy production". Namely, starting from (5.45), it can be confirmed directly that

$$\frac{d\mathcal{P}}{dt} = 2\frac{d\Psi}{dt} \leqq 0 \tag{5.55}$$

is valid, by which the correctness of the "dictionary" (5.41) as well as the validity of the Prigogine-Glansdorff theorem (5.25a) and (5.24) is proved [64, 8]. Cross-effects were not considered in the case of this simple example, and thus no use was made of the reciprocal relations in the proof of the theorems.

b) Stationary States of Thermodiffusional Systems with Chemical Reactions. Let us consider a multi-component fluid system without external forces ($\boldsymbol{F}_k = 0$), in which of K components R chemical reactions take place. Let us assume that in the sense determined by (2.118) the system is in a mechanical equilibrium. From the condition $\boldsymbol{F}_k = 0$, and from (2.118) we have that the equation of motion now has a particular form

$$\nabla p = 0 \tag{5.56}$$

hence, the pressure gradient vanishes in the system. In order to determine strictly the stationary states of purely dissipative processes with respect to heat conduction, thermodiffusion and chemical reactions, but without viscosity, it must be assumed that the barycentric velocity $\boldsymbol{v}$ can be neglected. According to the mass balance (2.17) this assumption means that the density of the continuum is considered as constant in time.

The system determined by the above conditions is characterized, apart from (5.56), by the component balances

$$\varrho\frac{\partial c_k}{\partial t} + \nabla\cdot\boldsymbol{J}_k = \sum_{j=1}^{R}\nu_{kj}J_j, \qquad (k = 1, 2, \ldots, K) \tag{5.57}$$

and by the energy balance

$$\varrho\frac{\partial u}{\partial t} + \nabla\cdot\boldsymbol{J}_{\mathrm{q}} = 0 \tag{5.58}$$

following from (2.46) and (3.36), further by the entropy production

$$\sigma = -\sum_{j=1}^{R}J_i\sum_{k=1}^{K-1}\nu_{kj}\frac{\mu_k}{T} - \sum_{k=1}^{K-1}\boldsymbol{J}_k\cdot\nabla\left(\frac{\mu_k-\mu_K}{T}\right) + \boldsymbol{J}_{\mathrm{q}}\cdot\nabla\left(\frac{1}{T}\right). \tag{5.59}$$

The latter is obtained from (3.87) disregarding the viscous terms.

The actual forms of the linear kinematical constitutive equations are obtained from (3.90), (3.92) and (3.93) with similar specifications

$$J_j = -\sum_{r=1}^{R} L_{jr} \sum_{k=1}^{K-1} \nu_{kr} \left(\frac{\mu_k - \mu_K}{T}\right), \qquad (j = 1, 2, \ldots, R) \tag{5.60}$$

$$\mathbf{J}_\mathrm{q} = L_\mathrm{qq} \nabla \left(\frac{1}{T}\right) - \sum_{k=1}^{K-1} L_{\mathrm{q}k} \nabla \left(\frac{\mu_k - \mu_K}{T}\right), \tag{5.61}$$

$$\mathbf{J}_i = L_{i\mathrm{q}} \nabla \left(\frac{1}{T}\right) - \sum_{k=1}^{K-1} L_{ik} \nabla \left(\frac{\mu_k - \mu_K}{T}\right), \quad (i = 1, 2, \ldots, K-1). \tag{5.62}$$

In the case of (5.60) with the validity of (2.38) for the affinities (3.70) it was taken into consideration that expression

$$A_r = -\sum_{k=1}^{K-1} \nu_{kr} \left(\frac{\mu_k - \mu_K}{T}\right), \qquad (r = 1, 2, \ldots, R) \tag{5.63}$$

holds. The reciprocal relations are

$$\begin{aligned} L_{jr} &= L_{rj}, \qquad (j, r = 1, 2, \ldots, R) \\ L_{\mathrm{q}k} &= L_{k\mathrm{q}}, \qquad (k = 1, 2, \ldots, K-1) \\ L_{ik} &= L_{ki}, \qquad (i, k = 1, 2, \ldots, K-1). \end{aligned} \tag{5.64}$$

Before proceeding, let us for the sake of simplification of the denotations, in particular for the sake of one of the generalizations to be carried out later on, let us introduce a set of new variables:

$$\Gamma_\mathrm{q} \equiv \frac{1}{T}, \qquad \Gamma_k \equiv -\frac{\mu_k - \mu_K}{T}, \qquad (k = 1, 2, \ldots, K-1). \tag{5.65}$$

By taking them into account in the above linear kinematical equations, the total entropy production of the system, by using the reciprocal relations as well, will be

$$\begin{aligned} \mathscr{P} = 2\Psi = \int_{V^0} \Big\{ & L_\mathrm{qq} (\nabla \Gamma_\mathrm{q})^2 + 2 \sum_{k=1}^{K-1} L_{\mathrm{q}k} \nabla \Gamma_\mathrm{q} \cdot \nabla \Gamma_k \\ & + \sum_{i,k=1}^{K-1} L_{ik} \nabla \Gamma_i \cdot \nabla \Gamma_k + \sum_{j,r=1}^{R} L_{jr} \sum_{i,k=1}^{K-1} \nu_{ij} \nu_{kr} \Gamma_i \Gamma_k \Big\} \, \mathrm{d}V^0 \end{aligned} \tag{5.66}$$

which is now the double of the adequate dissipation potential as well. By writing the condition of stationarity with the latter the variational task

$$\delta \Psi = 0 \tag{5.67}$$

is obtained. Carrying out the variation with respect to the parameters Γ_q and Γ_k, $(k = 1, 2, \ldots, K-1)$, the Euler-Lagrange equations of the problem

$$\frac{\partial \Psi}{\partial \Gamma_q} - \sum_{\alpha=1}^{3} \frac{\partial}{\partial x_\alpha} \frac{\partial \Psi}{\partial \left(\frac{\partial \Gamma_q}{\partial x_\alpha}\right)} = 0 \tag{5.68a}$$

and

$$\frac{\partial \Psi}{\partial \Gamma_i} - \sum_{\alpha=1}^{3} \frac{\partial}{\partial x_\alpha} \frac{\partial \Psi}{\partial \left(\frac{\partial \Gamma_i}{\partial x_\alpha}\right)} = 0, \qquad (i = 1, 2, \ldots, K-1) \tag{5.68b}$$

lead to the set of equations

$$L_{qq} \Delta \Gamma_q + \sum_{k=1}^{K-1} L_{qk} \Delta \Gamma_k = 0 \tag{5.69a}$$

and

$$L_{qi} \Delta \Gamma_q + \sum_{k=1}^{K-1} L_{ik} \Delta \Gamma_k - \sum_{j,r=1}^{R} L_{jr} \left(\sum_{k=1}^{K-1} \nu_{kr} \Gamma_k \right) \nu_{ij} = 0,$$
$$(i = 1, 2, \ldots, K-1) \tag{5.69b}$$

meanwhile the reciprocal relations (5.64) and the constancy of the coefficients have been used. With the linear kinematical constitutive equations (5.60), (5.61) and (5.62) the equations (5.69a) and (5.69b) can be written as

$$\nabla \cdot \boldsymbol{J}_q = 0, \tag{5.70a}$$

$$\nabla \cdot \boldsymbol{J}_i - \sum_{j=1}^{R} \nu_{ij} J_j = 0, \qquad (i = 1, 2, \ldots, K-1). \tag{5.70b}$$

With these equations obtained from the minimum principle and with the balance equations (5.57) and (5.58) the relations

$$\frac{\partial u}{\partial t} = 0 \tag{5.71a}$$

and

$$\frac{\partial c_i}{\partial t} = 0, \qquad (i = 1, 2, \ldots, K-1) \tag{5.71b}$$

expressing the stationarity are obtained. These are the direct conditions of the stationary energy and concentration distribution. Again it is proved by our results that the stationary states of the system are determined by the principle of least dissipation of energy and also by the principle of minimum production of entropy in the case of the validity of the linear Onsager theory.

Let us examine the stability of the above stationary state. By differentiating the dissipation function Ψ with respect to time and by using the (5.60), (5.61), (5.62) linear constitutive equations and the reciprocal relations (5.64), we get

$$\frac{\mathrm{d}\Psi}{\mathrm{d}t} = \int\limits_{V^0} \left\{ \boldsymbol{J}_\mathrm{q} \cdot \nabla \left(\frac{\partial \Gamma_\mathrm{q}}{\partial t} \right) + \sum_{k=1}^{K-1} \boldsymbol{J}_k \cdot \nabla \left(\frac{\partial \Gamma_k}{\partial t} \right) + \sum_{j=1}^{R} J_j \sum_{k=1}^{K-1} \nu_{kj} \frac{\partial \Gamma_k}{\partial t} \right\} \mathrm{d}V^0 \tag{5.72}$$

which by partial integration can be written as

$$\frac{\mathrm{d}\Psi}{\mathrm{d}t} = \int\limits_{\Omega^0} \left(\boldsymbol{J}_\mathrm{q} \frac{\partial \Gamma_\mathrm{q}}{\partial t} + \sum_{k=1}^{K-1} \boldsymbol{J}_k \frac{\partial \Gamma_k}{\partial t} \right) \cdot \mathrm{d}\boldsymbol{\Omega}^0$$

$$- \int\limits_{V^0} \left\{ \nabla \cdot \boldsymbol{J}_\mathrm{q} \frac{\partial \Gamma_\mathrm{q}}{\partial t} + \sum_{k=1}^{K-1} \left(\nabla \cdot \boldsymbol{J}_k - \sum_{j=1}^{R} \nu_{kj} J_j \right) \frac{\partial \Gamma_k}{\partial t} \right\} \mathrm{d}V^0 . \tag{5.73}$$

Since the values of the parameters Γ_q and Γ_k, $(k = 1, 2, \ldots, K - 1)$, i.e. the temperature T and the chemical potentials μ_k, $(k = 1, 2, \ldots, K - 1)$ are kept at constant values at the boundary, the surface integral in (5.73) vanishes, i.e.

$$\frac{\mathrm{d}\Psi}{\mathrm{d}t} = - \int\limits_{V^0} \left\{ \nabla \cdot \boldsymbol{J}_\mathrm{q} \frac{\partial}{\partial t} \frac{1}{T} - \sum_{k=1}^{K-1} \left(\nabla \cdot \boldsymbol{J}_k - \sum_{j=1}^{R} \nu_{kj} J_j \right) \frac{\partial}{\partial t} \frac{\mu_k - \mu_K}{T} \right\} \mathrm{d}V^0 \tag{5.74}$$

making use of the identities (5.65). Introducing the specific enthalpy

$$h = u + pv$$

we obtain the following energy balance from (5.58).

$$\varrho \frac{\partial h}{\partial t} + \nabla \cdot \boldsymbol{J}_\mathrm{q} = 0 , \tag{5.75}$$

since in this example the pressure is time-independent and the barycentric velocity was supposed to vanish, which implied the constancy of specific volume as well. With (5.57) and (5.75) the expression (5.74) may be given as

$$\frac{\mathrm{d}\Psi}{\mathrm{d}t} = \int\limits_{V^0} \left\{ \varrho \frac{\partial h}{\partial t} \frac{\partial}{\partial t} \frac{1}{T} - \sum_{k=1}^{K-1} \varrho \frac{\partial c_k}{\partial t} \frac{\partial}{\partial t} \frac{\mu_k - \mu_K}{T} \right\} \mathrm{d}V^0 . \tag{5.76}$$

According to the assumptions discussed in the above the specific enthalpy is a function of the state parameters, so that

$$h = h(T, c_1, c_2, \ldots, c_{K-1})$$

from which

$$\frac{\partial h}{\partial t} = \left(\frac{\partial h}{\partial T} \right)_{p, c_i} \frac{\partial T}{\partial t} + \sum_{k=1}^{K-1} \left(\frac{\partial h}{\partial c_k} \right)_{p, T, c_i} \frac{\partial c_k}{\partial t}$$

$$= c_\mathrm{p} \frac{\partial T}{\partial t} + \sum_{k=1}^{K-1} (h_k - h_K) \frac{\partial c_k}{\partial t} \tag{5.77}$$

is obtained. Here h_k is the partial specific enthalpy of the k-th component and c_p is the specific heat at constant pressure. By using the wellknown thermostatic relation

$$\mathrm{d}\left(\frac{\mu_k}{T}\right) = -\frac{h_k}{T^2}\,\mathrm{d}T + \frac{1}{T}\,(\mathrm{d}\mu_k)_T$$

in the modified form

$$\frac{\partial}{\partial t}\,\frac{\mu_k - \mu_K}{T} = -\frac{h_k - h_K}{T^2}\,\frac{\partial T}{\partial t} + \frac{1}{T}\sum_{i=1}^{K-1}\frac{\partial(\mu_k - \mu_K)}{\partial c_i}\,\frac{\partial c_i}{\partial t} \tag{5.78}$$

together with (5.77), we have

$$\frac{\mathrm{d}\Psi}{\mathrm{d}t} = -\int\limits_{V^0}\left\{\varrho\,\frac{c_\mathrm{p}}{T^2}\left(\frac{\partial T}{\partial t}\right)^2 + \frac{\varrho}{T}\sum_{i,k=1}^{K-1}\frac{\partial(\mu_k - \mu_K)}{\partial c_i}\,\frac{\partial c_i}{\partial t}\,\frac{\partial c_k}{\partial t}\right\}\mathrm{d}V^0. \tag{5.79}$$

Since ϱ, c_p and T are positive quantities, furthermore according to the stability condition

$$\sum_{i,k=1}^{K-1}\frac{\partial(\mu_k - \mu_K)}{\partial c_i}\,\mathrm{d}c_i\,\mathrm{d}c_k = \sum_{i,k=1}^{K-1}\frac{\partial^2 g}{\partial c_i\,\partial c_k}\,\mathrm{d}c_i\,\mathrm{d}c_k \geqq 0 \tag{5.80}$$

which is wellknown also from thermostatics [5] (where g is the specific Gibbs function), we can conclude that the expression (5.79) is always negative. Again this result means that during the evolution of the system towards the stationary state the dissipation potential Ψ always diminishes in the course of time, as long as the system did not attain a stationary state compatible with the imposed boundary conditions. In other words, the stationary state characterizable with the minimum of Ψ is stable. Of course, the same is valid for the entropy production $\mathscr{P} = 2\Psi$, hence (5.79) is a concrete proof for (5.29) and also for the Glansdorff-Prigogine theorem (5.24), owing to the validity of the basic postulates of the linear theory.

5. Generalizations

With the examination of the stability of the stationary heat conduction we started from (5.52), then for writing (5.53) the linear relation (5.43) was used. In the case of the second example we proceeded similarly. Thus (5.72) was obtained with the linear kinematical constitutive equations (5.60), (5.61) and (5.62) and the adequate Onsager's reciprocal relations. Briefly, in both cases, use was made of the basic postulates of the linear theory. This is plausible, since the Rayleigh-Onsager dissipation potential Ψ was defined for a linear theory only.

However, it can easily be seen that the right-hand sides of the expressions (5.53) and (5.72), which lead to the stability criteria of the stationary states, have a defined physical meaning also in the more general (non-linear) cases. Indeed, these expressions represent examples for the general relation (5.29) in linear as well as in non-linear cases, and thus the contributions of the change of the forces in time to the entropy production are always given by it. This means that the stability criteria (5.54) and (5.79) are of general validity, and are independent of the applicability of the linear theory, since the expressions (5.54), (5.79) and the general one (5.29) are also independent of the form of the kinematical constitutive equations. Hence, the entropy production due to the variation of forces is determined by expressions (5.53) and (5.72) for the case of non-linear phenomenological laws as well, but it is not true for the dissipation potential. Therefore, the dissipation potentials (4.9) and (4.10) defined only in the linear theory are not suitable for the treatment of non-linear cases, but due to the differential property of the entropy production, (5.29) can always be confirmed. However, unfortunately, nothing can be said in a direct manner about the variation of the entropy production (5.30) due to the variation of fluxes. Thus in each case, when (5.34a) and (5.34b) do not hold, the condition (5.36) of the minimum entropy production can not be confirmed either. The same refers to the validity of the "dictionary" (5.40) and (5.41) giving the relation between the Onsager and Prigogine principles in the linear case, i.e. it is not valid outside the domain of the linear theory.

As a matter of course, dissipation potentials can be defined in non-linear cases as well. Thus in some approximations a more detailed "dictionary" can be given between the variation of the non-linear dissipation potentials and the corresponding partial variations of the entropy production. A dissipation potential leading to a non-linear theory was first defined by Li [72]:

$$\Psi = \frac{1}{2} \sum_{i,k=1}^{f} L_{ik} X_i X_k + \frac{1}{6} \sum_{i,k,j=1}^{f} L_{ikj} X_i X_k X_j \tag{5.81}$$

where the quantities L_{ikj} are the second-order coefficients describing the non-linear effects. This dissipation potential (called "thermokinetic potential" by Li) is equivalent to the following non-linear kinematical constitutive equations

$$J_i = \sum_{k=1}^{f} L_{ik} X_k + \frac{1}{2} \sum_{k,j=1}^{f} L_{ikj} X_k X_j, \qquad (i = 1, 2, \dots, f) \tag{5.82}$$

furthermore with the reciprocal relations

$$L_{ik} = L_{ki}, \qquad (i, k = 1, 2, \dots, f) \tag{5.83}$$

and

$$L_{ikj} = L_{kij} = L_{kji} = \cdots, \qquad (i, k, j = 1, 2, \dots, f). \tag{5.84}$$

The non-linear theories are included in the above relations and have been suggested by Li [40, 72], Gyarmati [38, 51] and Rysselberghe

[39] independently from each other. It can be demonstrated that in the case of the dissipation potential (5.81) the principle of least dissipation of energy is valid [51]. Therefore, it is also certain that the relation between the Onsager and Prigogine principle can be extended to a non-linear region, where the relations (5.82), (5.83) and (5.84) are valid instead of the corresponding basic postulates of the linear theory.

Finally we refer to a new theory which has been developed by GLANSDORFF and PRIGOGINE in a way similar to the generalization of the principle of minimum production of entropy, which is also valid for cases when the conductivity coefficients are not constant, but functions of the local state parameters. In this theory the rôle of the entropy production is played by the so-called "local potentials", which may be considered similar to the Rayleigh-Onsager dissipation potentials in which the coefficients are the functions of the state parameters. However, though the use of the theory of "local potentials" has proved to be of real practical interest because it opens a way to use the wellknown variational techniques (Rayleigh-Ritz method, "self-consistent" method, etc.), this theory is not identical with a variational principle in the classical sense, but rather an extended variational technique involving functionals of two sets of functions.[1] Trial functions and iteration methods may be used in this theory. However, this theory and other non linear theories will not be dealt with, since at present these methods are being developed and cannot be regarded finished [67, 68, 69, 70, 71].

VI. The Integral Principle of Thermodynamics

Starting from the force representations of the principle of least dissipation of energy, first the Fourier equation of heat conduction is deduced in different pictures, then by generalizing the results the integral principle of thermodynamics is formulated (GYARMATI [50, 51, 53, 55, 73]). By repeating the method for the case of multi-component isothermal diffusion and viscous flow, the Fick equations (VERHÁS [60, 74]) and the Navier-Stokes equation is derived in the general form (VERHÁS [60, 74], BÖRÖCZ [75]).

It is shown by the applications mentioned that the force representation of the principle of least dissipation of energy is more productive than the flux representation, further, that the Euler-Lagrange equations belonging to the integral principle are equivalent to the total set of the irreversible transport equations. For the sake of its direct conformation and as an application of the integral principle, the deduction

[1] These remarks were originally made by PRIGOGINE and GLANSDORFF [71], but we agree in every respect.

of non-isothermal transport equations, in which cross-effects between the phenomena occur, will be dealt with (VERHÁS [76]). In the following a general form of the transport equations will be given by starting from the force representation of the principle of least dissipation of energy (GYARMATI). This deduction points out the intrinsic connection between the integral principle and the principle of least dissipation of energy, more precise its force representation, in a general manner. The relation between the Hamilton principle and the thermodynamic integral principle is discussed (GYARMATI [73]), and the canonical field equations belonging to the integral principle of thermodynamics are determined (VERHÁS [78], VOJTA [79]). Finally, the Legendre transformations of the dissipation potentials and of the dissipative Lagrange and Hamilton densities are treated and, the canonical form of the dissipative integral is given (GYARMATI).

1. Deduction of the Fourier Equation

It has already been mentioned several times that, though the flux and force representations of the principle of least dissipation of energy are in principle equivalent to each other, in practical respects the situation is different. In the case of the formulation of the principle in flux representation, where the stipulation that the forces must be kept at a constant value in the course of the variation with respect to fluxes the direct deduction of the transport equations (the Fourier, Fick, Navier-Stokes etc. equations) is a priori rendered impossible. The reason for this is that for the deduction of the transport equations governing heat conduction, diffusion, viscous flow, etc. it is necessary that the temperature, chemical potentials, velocities etc., i.e. in general, the intensive quantities must be varied. This fact, however, is incompatible with the flux representation, where the constancy of the forces determined by the negative gradients of the intensive quantities is required. This difficulty is automatically eliminated in the force representation, and consequently, it is justified to expect that the force representation will be more productive (at least in practical respect just as in the stationary case) than the flux representation.

In the following by starting from the general global form of the force representation

$$\delta \int_V [\varrho \dot{s} + \nabla \cdot \boldsymbol{J}_s - \Psi]_J \, \mathrm{d}V = \delta \int_V [\sigma - \Psi]_J \, \mathrm{d}V = 0, \qquad (6.1)$$

the transport equations are deduced one after the other and the thermodynamic integral principle is formulated. The derivation of the transport equations will start with one of the most delicate ones: the deduction of the Fourier equation of heat conduction. This will be carried out (in the case of the heat conduction of solid bodies) in three different actual pictures and then a quite general one will be given, since, the comparison of the deductions in the different pictures is very instructive.

The basis for the deduction in three different pictures is given by the fact that, the heat current density can usually be given by three different coefficients and forces. Namely

$$\boldsymbol{J}_{\mathrm{q}} = -\lambda \nabla T = -L^*_{\mathrm{qq}} \nabla \ln T = L_{\mathrm{qq}} \nabla \frac{1}{T}, \tag{6.2}$$

where the forces correspond to the choice

$$\boldsymbol{X}^{**}_{\mathrm{q}} \equiv -\nabla T, \quad \boldsymbol{X}^*_{\mathrm{q}} \equiv -\frac{\nabla T}{T} = -\nabla \ln T, \quad \boldsymbol{X}_{\mathrm{q}} \equiv \nabla \frac{1}{T}, \tag{6.3}$$

whereas between the coefficients the relations

$$\lambda = T^{-1} L^*_{\mathrm{qq}} = T^{-2} L_{\mathrm{qq}} \tag{6.4}$$

are valid. Of course, the entropy production belonging to the heat conduction can also be given in three forms, i.e.

$$\sigma_{\mathrm{q}} = \frac{\boldsymbol{J}_{\mathrm{q}} \cdot \boldsymbol{X}^{**}_{\mathrm{q}}}{T^2} = \frac{\boldsymbol{J}_{\mathrm{q}} \cdot \boldsymbol{X}^*_{\mathrm{q}}}{T} = \boldsymbol{J}_{\mathrm{q}} \cdot \boldsymbol{X}_{\mathrm{q}}, \tag{6.5}$$

moreover, also three alternative local dissipation potentials can be defined

$$\Psi^{**}_{\mathrm{q}} \equiv \frac{\lambda}{2} (\nabla T)^2, \quad \Psi^*_{\mathrm{q}} \equiv \frac{L^*_{\mathrm{qq}}}{2} (\nabla \ln T)^2, \quad \Psi_{\mathrm{q}} \equiv \frac{L_{\mathrm{qq}}}{2} \left(\nabla \frac{1}{T}\right)^2 \tag{6.6}$$

between which the relations

$$\Psi^{**}_{\mathrm{q}} = T \Psi^*_{\mathrm{q}} = T^2 \Psi_{\mathrm{q}} \tag{6.7}$$

are valid. By proceeding in the given expressions from left to right, the quantities in question are given in Fourier's picture (λ and the quantities which are denoted by **), in energy picture (quantities are denoted by *), and in entropy picture (the quantities not denoted by asterisks), respectively. Primarily the heat conduction equation of solid bodies is derived in Fourier's picture. It will be seen that, in this representation the original form of the equation is directly obtained, but it becomes also evident, that from the theoretical viewpoint this is not a very lucky representation.

a) The Fourier Picture. In the case of pure heat conduction one has to start from the particular form

$$\varrho \dot{s} + \nabla \cdot \left(\frac{\boldsymbol{J}_{\mathrm{q}}}{T}\right) = \sigma_{\mathrm{q}}, \qquad \boldsymbol{J}_{\mathrm{s}} = \frac{\boldsymbol{J}_{\mathrm{q}}}{T}, \tag{6.8}$$

of the general entropy balance (3.60). On the other hand, the formulation of the variational principle (6.1) in Fourier's picture means that the variational condition must directly be given with the dissipation potential Ψ_q^{**}, i.e. the variational condition

$$\delta\int\limits_V \left\{T^2\left[\varrho\dot{s} + \nabla\cdot\left(\frac{J_q}{T}\right)\right] - \Psi_q^{**}\right\}_{J_q} \mathrm{d}V = \delta\int\limits_V [T^2\sigma_q - \Psi_q^{**}]_{J_q}\,\mathrm{d}V = 0 \tag{6.9}$$

shall be considered. It is needless to emphasize that this form of the variational principle is not the most desirable one, owing to the multiplication of the local form by T^2, at least from the aesthetical point of view. However, it is Fourier who is responsible for this fact who considered the negative gradient of temperature a thermal force according to the first expression of (6.2).

Taking into account that in the case of a solid body

$$\dot{s} = \frac{1}{T}\dot{u} = \frac{c_v}{T}\frac{\partial T}{\partial t} \tag{6.10}$$

is valid, where c_v is the specific heat at constant volume, on the basis of (3.58), with the expressions (6.6) and (6.10) from the variational condition (6.9) the extremum principle

$$L = \int\limits_V \left[\varrho c_v T\frac{\partial T}{\partial t} + T\,\nabla\cdot\boldsymbol{J}_q - \boldsymbol{J}_q\cdot\nabla T - \frac{\lambda}{2}(\nabla T)^2\right]_{J_q} \mathrm{d}V = \max \tag{6.11}$$

is obtained, where the integral is considered a function of temperature alone. Hence, in (6.11) we have to vary with respect to the internal variable of the force $\boldsymbol{X}_q^{**} \equiv -\nabla T$, i.e. with respect to alone the temperature at constant flux. However, according to the internal energy balance

$$\varrho\frac{\partial u}{\partial t} + \nabla\cdot\boldsymbol{J}_q = \varrho c_v\frac{\partial T}{\partial t} + \nabla\cdot\boldsymbol{J}_q = 0 \tag{6.12}$$

obtained by specialization from (3.37) it is plausible that from the non-variation of the flux: $\delta\boldsymbol{J}_q = 0$, the non-variation of $\varrho\frac{\partial u}{\partial t}$ $\left(\text{or } \varrho c_v\frac{\partial T}{\partial t}\right)$ follows as well. Therefore, varying (6.11) with respect to temperature alone, in case of fixed $\boldsymbol{J}_q$ and $\varrho c_v\frac{\partial T}{\partial t}$, we get

$$\delta L = \int\limits_V \left[\left(\varrho c_v\frac{\partial T}{\partial t} + \nabla\cdot\boldsymbol{J}_q\right)\delta T - (\boldsymbol{J}_q + \lambda\nabla T)\cdot\nabla\,\delta T\right]\mathrm{d}V = 0 \tag{6.13}$$

which is the necessary condition of the maximum (6.11). This relation will be reduced by the first form of the linear laws (6.2) to the condition

$$\delta L = \int\limits_V \left[\varrho c_v\frac{\partial T}{\partial t} - \nabla\cdot(\lambda\nabla T)\right]\delta T\,\mathrm{d}V = 0\,. \tag{6.14}$$

Since the local δT variation is arbitrary, the maximum principle (6.11) is equivalent to the existence of the Fourier equation

$$\varrho c_{\mathrm{v}} \frac{\partial T}{\partial t} = \nabla \cdot (\lambda \nabla T). \tag{6.15}$$

It should be emphasized that in the course of the deduction no use was made of the constancy in space of the heat conductivity coefficient, hence λ may depend on the space coordinates.

In the preceedings our attention was focused on the direct deduction of the Fourier differential equation. Now, the Lagrange function belonging to the variational principle will be determined by variation along the boundary surfaces of the system. For this purpose let us transform the second term of the extremum condition (6.14) in the following way:

$$-\delta T\, \nabla \cdot (\lambda\, \nabla T) = -\nabla \cdot (\delta T\, \lambda \nabla\, T) + \lambda\, \nabla T \cdot \nabla\, \delta T. \tag{6.16}$$

Since the Gauss theorem for the divergence term can be applied, i.e.

$$\int\limits_V \nabla \cdot (\delta T\, \lambda\, \nabla T)\, \mathrm{d}V = \oint\limits_\Omega \lambda\, \delta T \left(\frac{\partial T}{\partial \boldsymbol{n}}\right) \cdot \mathrm{d}\boldsymbol{\Omega} \tag{6.17}$$

(where $\boldsymbol{n}$ is the outward normal to Ω) furthermore

$$\lambda\, \nabla T \cdot \nabla\, \delta T = \delta \left[\frac{\lambda}{2} (\nabla T)^2\right] \tag{6.18}$$

is also valid, and consequently the extremum condition can be given in the following form as well:

$$\delta L = \int\limits_V \left\{\delta \left(\varrho c_{\mathrm{v}} T \frac{\partial T}{\partial t}\right) + \delta \frac{\lambda}{2} (\nabla T)^2\right]\Big\} \mathrm{d}V - \oint\limits_\Omega \lambda\, \delta T \left(\frac{\partial T}{\partial \boldsymbol{n}}\right) \cdot \mathrm{d}\boldsymbol{\Omega} = 0. \tag{6.19}$$

If we agree that the temperature along the boundary surface will not be varied, the necessary condition of the maximum (6.11) will be determined by the volume integral of (6.19) already:

$$\delta L = \delta \int\limits_V \left[\varrho c_{\mathrm{v}} T \frac{\partial T}{\partial t} + \frac{\lambda}{2} (\nabla T)^2\right] \mathrm{d}V = 0. \tag{6.20}$$

From this form of the necessary condition of the variational task already it is seen that the Lagrangian density of the variational problem referring to heat conduction is:

$$\mathscr{L}_T = \mathscr{L}_T(T, \nabla T) = \varrho c_{\mathrm{v}} T \frac{\partial T}{\partial t} + \frac{\lambda}{2} (\nabla T)^2. \tag{6.21}$$

Here the varied parameter was denoted as an index, and we emphasize that $\mathscr{L}_T$ depends on $\varrho c_{\mathrm{v}} \frac{\partial T}{\partial t}$ as a constant quantity, i.e. the time derivative $\frac{\partial T}{\partial t}$ is not a new independent variable of $\mathscr{L}_T$ owing to the energy balance (6.12). With this Lagrangian density the following variational task

$$\delta L[T] = \delta \int_V \mathscr{L}_T \, \mathrm{d}V = 0 \tag{6.22}$$

leads to the Euler-Lagrange equation

$$\frac{\partial \mathscr{L}_T}{\partial T} - \sum_{\alpha=1}^{3} \frac{\partial}{\partial x_\alpha} \frac{\partial \mathscr{L}_T}{\partial \left(\frac{\partial T}{\partial x_\alpha}\right)} = 0 \tag{6.23}$$

which is identical to (6.15), the Fourier equation. It should be noted that the above deduction of the Fourier equation was essentially a sufficient basis for the general formulation of the integral principle of thermodynamics [50, 51]. However, before formulating the integral principle in a general form, we want to deduce the equation of heat conduction in energy and in entropy representations [53], as well as in a generalized "Γ" picture [80].

b) The Energy Picture. In case of the energy picture let us start from the central expression of the entropy production (6.5) by which and with Ψ_q^* defined in (6.6), the variational condition (6.1) can be given in the form

$$\delta \int_V \left\{ T\left[\varrho \dot{s} + \nabla \cdot \left(\frac{\boldsymbol{J}_q}{T}\right)\right] - \Psi_q^* \right\} \mathrm{d}V = \delta \int_V [T\sigma_q - \Psi_q^*] \, \mathrm{d}V = 0. \tag{6.24}$$

Let us consider the right-hand side formulation which was particularly used by Verhás [60, 74], and which leads with the central expressions of (6.2) and (6.6) to the extremum principle

$$-\int_V \left[\boldsymbol{J}_q \cdot \nabla \ln T + \frac{L_{qq}^*}{2} (\nabla \ln T)^2\right] \mathrm{d}V = \max. \tag{6.25}$$

This form of the principle may be rewritten with the identity

$$\nabla \cdot (\boldsymbol{J}_q \ln T) = \boldsymbol{J}_q \cdot \nabla \ln T + \ln T \, \nabla \cdot \boldsymbol{J}_q \tag{6.26}$$

and by the energy balance (6.12) in the form

$$-\int_V \left[\varrho c_v \ln T \frac{\partial T}{\partial t} + \frac{L_{qq}^*}{2} (\nabla \ln T)^2\right] \mathrm{d}V - \oint_\Omega \boldsymbol{J}_q \ln T \cdot \mathrm{d}\boldsymbol{\Omega} = \max \tag{6.27}$$

while the volume integral of the divergence term $\nabla \cdot (\boldsymbol{J}_q \ln T)$ has been transformed with Gauss' theorem into a surface integral. Let the maximum (6.27) be considered at similar variational conditions as previously in the case of the Fourier picture. If we agree that we vary with respect to the temperature (more precise $\ln T$) alone a fixed $\varrho c_v \frac{\partial T}{\partial t}$ and $\boldsymbol{J}_q$, furthermore that the temperature will not be varied at the boundaries the maximum (6.27) is already determined by the volume integral, i.e.

$$-\int_V \left[\varrho c_V \ln T \frac{\partial T}{\partial t} + \frac{L^*_{qq}}{2} (\nabla \ln T)^2\right] dV = \max. \qquad (6.28)$$

Since in (6.27) the heat current is contained in the surface integral only, it can be briefly said that the extremum principle (6.27) is valid at fixed $\varrho c_V \frac{\partial T}{\partial t}$ and in the case, if nothing is varied along the boundary surface. From (6.28) the following expression is obtained for the Lagrangian density of the variational problem:

$$\mathscr{L}_{\ln T} = \mathscr{L}_{\ln T} (\ln T, \nabla \ln T) = -\varrho c_V \ln T \frac{\partial T}{\partial t} - \frac{L^*_{qq}}{2} (\nabla \ln T)^2. \qquad (6.29)$$

With this Lagrangian density, the integral principle can be formulated in a concise form as

$$\delta L[\ln T] = \delta \int_V \mathscr{L}_{\ln T} \, dV = 0, \qquad (6.30)$$

to which the following Euler-Lagrange equation belongs:

$$\frac{\partial \mathscr{L}_{\ln T}}{\partial \ln T} - \sum_{\alpha=1}^{3} \frac{\partial}{\partial x_\alpha} \frac{\partial \mathscr{L}_{\ln T}}{\partial \left(\frac{\partial \ln T}{\partial x_\alpha}\right)} = 0. \qquad (6.31)$$

By carrying out the indicated differentiation the equation

$$\varrho c_V \frac{\partial T}{\partial t} = -\nabla \cdot (L^*_{qq} \nabla \ln T) \qquad (6.32)$$

is obtained, which can be rewritten in the wellkown form of the Fourier equation (6.15), by using the central expressions of (6.3) and (6.4). It should be noted that L^*_{qq} may be a function of the coordinates further that our above deductions are not at all modified, if λ and L^*_{qq} are tensor quantities, i.e. if heat conduction in anisotropic solid bodies is concerned.

c) The Entropy Picture. The deduction of the Fourier equation in the entropy picture can be carried out very briefly. Not only because several hitherto mentioned details can be disregarded, but rather because this picture corresponds without any modification to the variational condition formulated in (6.1). Indeed, by rewriting the left-hand side of (6.1) with the use of the dissipation potential Ψ_q of (6.6) and after applying the divergence theorem, we have

$$\delta \int_V [\varrho \dot{s} - \Psi_q] \, dV + \delta \oint_\Omega \boldsymbol{J}_s \cdot d\boldsymbol{\Omega} = 0. \qquad (6.33)$$

If we agree that nothing will be varied at the boundaries (i.e. $\boldsymbol{J}_s$, or $\boldsymbol{J}_q$ and T are fixed at the boundaries with respect to the variation)[1], with the form (6.6) of Ψ_q and with (6.10) the variational condition

$$\delta \int\limits_V \left[\varrho \frac{c_v}{T} \frac{\partial T}{\partial t} - \frac{L_{qq}}{2} \left(\nabla \frac{1}{T}\right)^2\right] dV = 0 \tag{6.34}$$

is obtained, where the variation must be taken exclusively with respect to $\frac{1}{T}$ at fixed $\varrho c_v \frac{\partial T}{\partial t}$. Proceeding in this way, we get

$$\int\limits_V \left\{\left[\varrho c_v \frac{\partial T}{\partial t} + \nabla \cdot \left(L_{qq} \nabla \frac{1}{T}\right)\right] \delta \left(\frac{1}{T}\right)\right\} dV = 0 \tag{6.35}$$

from which the differential equation

$$\varrho c_v \frac{\partial T}{\partial t} = -\nabla \cdot \left(L_{qq} \nabla \frac{1}{T}\right) \tag{6.36}$$

follows as Euler-Lagrange equation. This differential equation is the equation of heat conduction in the entropy picture, since by using the last expressions of (6.3) and (6.4) it can be rewritten in the wellknown form (6.15) of the Fourier equation.

From the variational condition (6.34) the Lagrangian density can be read off:

$$\mathscr{L}_{\frac{1}{T}} = \mathscr{L}_{\frac{1}{T}}\left(\frac{1}{T}, \nabla \frac{1}{T}\right) = \varrho \frac{c_v}{T} \frac{\partial T}{\partial t} - \frac{L_{qq}}{2} \left(\nabla \frac{1}{T}\right)^2 \tag{6.37}$$

where the index of the Lagrangian density again refers to the varied parameter. With this Lagrangian density the equation (6.36) can be considered as the Euler-Lagrange equation of the variational task

$$\delta L\left[\frac{1}{T}\right] = \delta \int\limits_V \mathscr{L}_{\frac{1}{T}}\, dV = 0 \tag{6.38}$$

since, the equation

$$\frac{\partial \mathscr{L}_{\frac{1}{T}}}{\partial \frac{1}{T}} - \sum_{\alpha=1}^{3} \frac{\partial}{\partial x_\alpha} \frac{\partial \mathscr{L}_{\frac{1}{T}}}{\partial \left(\dfrac{\partial \frac{1}{T}}{\partial x_\alpha}\right)} = 0 \tag{6.39}$$

belongs to the condition (6.38).

Summarizing it can be said that heat conduction in solid bodies is described by equations (6.15), (6.32) and (6.36), and so they are practically equivalent to each other. Theoretically, however, the energy representation, and rather the Fourier picture-leading directly to the traditional form (6.15) of the equation of heat conduction are un-

[1] Evidently, the above and also the previously presented variational formulations remain valid for "free boundary conditions" as well.

fortunate, since in these cases, the general variational principle (6.1) must be modified by the multiplicator T and T^2, respectively. Though the use of the energy picture is advantageous particularly in the case of isothermal problems, in non-isothermal systems it is generally preferable to use the entropy picture for the description of non-isothermal transport processes.

d) The Generalized "Γ" Picture. The derivation of the equation of heat conduction in the previous three pictures corresponds to the cases permitted by the (6.2) forms of Fourier's linear law usually quoted in irreversible thermodynamics. Now, generalizing the foregoing, we give the derivation of the Fourier equation in a universal "Γ" picture containing the results obtained in the entropy, the energy and the Fourier pictures as special cases (FARKAS [80]).

Considering an arbitrary $\Gamma = f(T)$ function of the absolute temperature interpreted in the interval

$$0 < T < \infty$$

whose derivatives with respect to temperature fulfil the condition

$$-\infty < f'(T) < 0 \tag{6.40}$$

in the total open interval $(0, \infty)$. Now let us introduce a universal "Γ" picture, in which the thermal force will be

$$\boldsymbol{X}_\Gamma \equiv \nabla \Gamma = f'(T) \nabla T, \tag{6.41}$$

whereas the L_Γ phenomenological coefficient, which is constant in this "Γ" picture, is defined by Fourier's linear law

$$\boldsymbol{J}_{\mathrm{q}} = L_\Gamma \nabla \Gamma = -\lambda \nabla T, \tag{6.42}$$

since the $\boldsymbol{J}_{\mathrm{q}}$ heat flow must be the same in every representation. Hence follows that

$$L_\Gamma = -\frac{\lambda}{f'(T)} \tag{6.43}$$

with the condition

$$0 < \lambda < \infty$$

is valid for the common heat conductivity coefficient and similarly (6.40) leads to the fulfilment of

$$0 < L_\Gamma < \infty .$$

The quantities

$$\sigma_\Gamma(\boldsymbol{J}_{\mathrm{q}}, \boldsymbol{X}_\Gamma) = \boldsymbol{J}_{\mathrm{q}} \cdot \boldsymbol{X}_\Gamma = \boldsymbol{J}_{\mathrm{q}} \cdot \nabla \Gamma \tag{6.44}$$

and

$$\Psi_\Gamma(\boldsymbol{X}_\Gamma, \boldsymbol{X}_\Gamma) = \frac{L_\Gamma}{2} \boldsymbol{X}_\Gamma^2 = \frac{L_\Gamma}{2} (\nabla \Gamma)^2 \tag{6.45}$$

corresponding to the local entropy production and the dissipation potential in the "Γ" picture, of course, are not the quantities of the dimension of the entropy production in a general case.

Now let us start from the form of the least dissipation of energy

$$\delta \int_V [\sigma_\Gamma - \Psi_\Gamma] \, \mathrm{d}V = \delta \int_V \left[\boldsymbol{J}_\mathrm{q} \cdot \nabla \Gamma - \frac{L_\Gamma}{2} (\nabla \Gamma)^2 \right] \mathrm{d}V = 0 \quad (6.46)$$

formulated with the above introduced quantities. This can be rewritten by using the identity

$$\boldsymbol{J}_\mathrm{q} \cdot \nabla \Gamma = \nabla \cdot (\boldsymbol{J}_\mathrm{q} \Gamma) - \Gamma \, \nabla \cdot \boldsymbol{J}_\mathrm{q}$$

and the Gauss theorem

$$\int_V \nabla \cdot (\boldsymbol{J}_\mathrm{q} \Gamma) \, \mathrm{d}V = \oint_\Omega (\boldsymbol{J}_\mathrm{q} \Gamma) \cdot \mathrm{d}\boldsymbol{\Omega}$$

in the following form

$$\delta \int_V \left[-\Gamma \, \nabla \cdot \boldsymbol{J}_\mathrm{q} - \frac{L_\Gamma}{2} (\nabla \Gamma)^2 \right] \mathrm{d}V = 0 \quad (6.47)$$

keeping in mind the auxiliary condition of variation according to which we do not vary the current densities. According to the energy balance (6.12), this prescription is equivalent to the condition

$$\delta \left(\varrho \frac{\partial u}{\partial t} \right) = \delta \left(\varrho c_\mathrm{v} \frac{\partial T}{\partial t} \right) = 0, \quad (6.48)$$

i.e. the quantity $\varrho c_\mathrm{v} \frac{\partial T}{\partial t}$ is considered as constant during the variation. Of course, $\delta L_\Gamma = 0$, i.e. L_Γ is constant, and moreover, Γ is also fixed at the boundaries. The constancy of L_Γ corresponds to the premise of Onsager's linear theory operating with constant phenomenological coefficients, whereas the prescription $\delta \Gamma = 0$ on the boundary surface is equivalent to $\delta T = 0$ also at the boundaries owing to (6.40). With the above-mentioned conditions and by using (6.12), formula (6.47) leads to the variational task

$$\delta \int_V \left[\Gamma \varrho c_\mathrm{v} \frac{\partial T}{\partial t} - \frac{L_\Gamma}{2} (\nabla \Gamma)^2 \right] \mathrm{d}V = 0$$

to which the following differential equation

$$\varrho c_\mathrm{v} \frac{\partial T}{\partial t} + \nabla \cdot (L_\Gamma \nabla \Gamma) = 0 \quad (6.49)$$

belongs as an Euler-Lagrange equation. This differential equation describes the heat conduction in solids in an universal form. Accordingly, (6.49) includes (6.36) of the entropy picture in the case of $\Gamma \equiv \frac{1}{T}$,

(6.32) of the energy picture in the case of $\Gamma \equiv -\ln T$, and finally the traditional (6.15) form of Fourier's equation if $\Gamma \equiv -T$. We emphasize again that the local quantities σ_Γ and Ψ_Γ defined in the equations (6.44) and (6.45) have the dimension of the entropy production only if $\Gamma \equiv \frac{1}{T}$ and $L_\Gamma \equiv L_{qq}$ that is, in the case of the entropy picture. This is evident from expressions (6.5) and (6.6) underlying the three representations discussed in detail.

The way of description in the universal "Γ" picture is of great importance not only because it contains the former results of the different pictures in an essential form, offering thereby a uniform way of theoretical description, but it offers well-based hopes in practical respect as well. For example— with some stipulations depending on the actual problems—it allows the treating problems of heat conduction in a simple way, which could only be described with the non-linear equation

$$\varrho c_v \frac{\partial T}{\partial t} = \nabla \cdot [\lambda(T) \nabla T]$$

because of the dependence on temperature of the heat-conductivity coefficient $\lambda = \lambda(T)$. If we succeed in establishing the $\lambda(T)$ function (for instance experimentally in actual cases), we can seek for such a "Γ" picture, the coefficient L_Γ of which is constant according to (6.43). In such cases the non-linear variational task in the universal "Γ" picture is reduced to a linear variational problem.

Naturally the non-linear variational tasks, valid in the dependence of λ on temperature and leading to the non-linear form of the transport-equations of the type similar to the last equation, represent a type of non-linearity very different from that, a possible theory of which was mentioned in Chapter V. sect. 5. This is evident, because the non-linear constitutive equations in (5.82) define non-linear relations between fluxes and forces, and if this is called non-linearity in an exact sense, we have to say that by treating the problem in the universal "Γ" picture only the $\lambda = \lambda(T)$ (and other similar) non-linearities, i.e. non-linearities in a weaker sense can be eliminated. Of course, the introduction of a universal "Γ" picture is possible not only in the case of heat-conduction, but—by the fulfilment of suitable conditions—in the case of other transport problems as well. This is why the universal "Γ" picture worked out by Farkas [80], is very useful in practical respects.

2. Formulation of the Integral Principle

In the preceding part a new integral principle was formulated by starting from the global form (6.1) of the principle of least dissipation of energy given in the force representation for the particular case of heat conduction. This principle is expressed in different pictures by the variational conditions (6.22), (6.30), (6.38) and (6.47). Let us consider the form of the Lagrangian density given in detail in (6.37) pertaining to the integral principle formulated in the entropy picture (6.38) which can be given in the concise form

$$\mathscr{L}_{\frac{1}{T}} = \varrho \dot{s} - \Psi_{\mathrm{q}} \tag{6.50}$$

on the basis of the last expression of (6.6) and (6.10). By comparing it to (6.33), serving as the starting point, it can be seen that theLagrangian density is identical to the integrand of the volume integral (6.33). In (6.33) the surface integral has an influence only upon the boundary conditions (its vanishing means that $\boldsymbol{J}_{\mathrm{s}}$ or $\boldsymbol{J}_{\mathrm{q}}$ and T are fixed at the boundaries during the δ-process) and consequently, the boundary conditions in the form (6.50) of the Lagrangian density are already taken into account. The same is evidently also valid for the form of integral principle (6.38).

Now our results will be generalized considering the conditions stated for the case of heat conduction and in this way the integral principle of thermodynamics will be formulated. Since $\dot{s}$ can always be given with the relation (3.25), following from the Gibbs relation for the case of systems in cellular equilibrium, furthermore the dissipation potential Ψ can also be taken as known, it is evident that in the entropy picture the function

$$\mathscr{L} = \varrho \dot{s} - \Psi \tag{6.51a}$$

or more generally the following function

$$\mathscr{L} = \sigma - \Psi \tag{6.51b}$$

must be considered as the Lagrangian density of thermodynamics. This can be expressed in such a manner that the thermodynamic Lagrangian density is the difference of the "kinetic" part $\varrho \dot{s}$ containing the time derivative and of the "potential" part determined by the dissipation potential Ψ.

Now let us denote the intensive parameters which occur in the entropy picture (3.25), and the gradients of which are the corresponding $\boldsymbol{X}_i$ forces by Γ_i. Then in the case of f independent scalar parameters Γ_i for the global L Lagrange function, considered as their functional, the integral principle

$$L = \int_V \mathscr{L} \, \mathrm{d}V = \max \tag{6.52}$$

or, the variational condition

$$\delta L[\Gamma_1, \Gamma_2, \ldots, \Gamma_f] = \delta \int_V \mathscr{L} \, \mathrm{d}V = 0 \tag{6.53}$$

equivalent to it is valid, to which the Euler-Lagrange equations

$$\frac{\partial \mathcal{L}}{\partial \Gamma_i} - \sum_{\alpha=1}^{3} \frac{\partial}{\partial x_\alpha} \frac{\partial \mathcal{L}}{\partial \left(\frac{\partial \Gamma_i}{\partial x_\alpha}\right)} = 0, \quad (i = 1, 2, \ldots, f) \tag{6.54}$$

belong. In the case of the variational task (6.53) the variation must be exclusively taken with respect to the Γ_i parameters, the gradients of which determine the forces. This prescription of variation can be understood, if it is also taken into consideration that the integral principle formulated in the form (6.52) or (6.53) is in a very intrinsic relation with the force representation of the principle of least dissipation of energy, since it refers to parameters the gradients of which are the forces. In the course of the formulation of the principle of least dissipation of energy in the force representation [see (4.21) to (4.25)], the variation was carried out only with respect to the forces at constant fluxes. Since owing to the balance equations, the time derivatives of the parameters occurring in them (and also in $\dot{s}$) are determined by the fluxes [see, for instance (6.12)], the conditions, according to which the time derivatives are not varied, is essentially a consequence of the non-variation of the fluxes. However, the latter variational convention is not a new and independent one, because of this being a property of the Onsager principle formulated in the force representation.

Before analysing further the theoretical problems it will be proved with examples that, the principle formulated in general by the relations (6.51) to (6.54) is indeed the genuine and exact integral principle of thermodynamics. This is equivalent to the total set of transport equations governing the evolution of irreversible processes in space-time.

3. Deduction of the Fick Equations of Isothermal Diffusion

Let us consider a continuum of K components in absence of external forces in which temperature and pressure are uniform, but the gradients of the chemical potentials do not vanish. If for the moment the chemical reactions are disregarded, in a system satisfying the above conditions only diffusion processes take place, which result locally in the energy dissipation

$$T\sigma_{\mathrm{d}} = \sum_{k=1}^{K-1} \boldsymbol{J}_k \cdot \boldsymbol{X}_k^* \tag{6.55}$$

where the quantities

$$\boldsymbol{X}_k^* \equiv -\nabla(\mu_k - \mu_K), \quad (k = 1, 2, \ldots, K-1) \tag{6.56}$$

are the forces of diffusion. The above expressions are given in the energy picture, advantageous for the case of isothermal problems and, following from the general equation (3.87). It is preferable to introduce the independent parameters

$$\Gamma_k^* \equiv -(\mu_k - \mu_K), \qquad (k = 1, 2, \ldots, K-1) \tag{6.57}$$

by which the dissipation potential, in energy picture, owing to (4.15a) will be

$$\Psi_{\mathrm{d}}^* \equiv \frac{1}{2} \sum_{i,k=1}^{K-1} L_{ik}^* \boldsymbol{X}_i^* \cdot \boldsymbol{X}_k^* = \frac{1}{2} \sum_{i,k=1}^{K-1} L_{ik}^* \nabla \Gamma_i^* \cdot \nabla \Gamma_k^* . \tag{6.58}$$

Now following VERHÁS [60, 74], we write the right-hand side of the principle (6.1) of least dissipation of energy in energy picture, i.e. we start from the following variational condition

$$\delta \int_V [T\sigma_{\mathrm{d}} - \Psi_{\mathrm{d}}^*] \, \mathrm{d}V = 0 \tag{6.59}$$

to which, by using (6.55) and (6.58) the extremum principle

$$\int_V \left[\sum_{k=1}^{K-1} \boldsymbol{J}_k \cdot \nabla \Gamma_k^* - \frac{1}{2} \sum_{i,k=1}^{K-1} L_{ik}^* \nabla \Gamma_i^* \cdot \nabla \Gamma_k^* \right] \mathrm{d}V = \max \tag{6.60}$$

belongs. (It should be noted that by changing the sign also minimum could be written [60, 74], however, for the sake of uniform treatment the maximum formulation always used hitherto will be maintained.) By making use of the vector analytical identity

$$\nabla \cdot (\boldsymbol{J}_k \Gamma_k^*) = \Gamma_k^* \nabla \cdot \boldsymbol{J}_k + \boldsymbol{J}_k \cdot \nabla \Gamma_k^*, \qquad (k = 1, 2, \ldots, K-1) \tag{6.61}$$

and by transforming the divergencies $\int_V \nabla \cdot (\boldsymbol{J}_k \Gamma_k^*) \, \mathrm{d}V$ with the Gauss theorem into surface integrals, we get

$$-\int_V \left[\sum_{k=1}^{K-1} \Gamma_k^* \nabla \cdot \boldsymbol{J}_k + \frac{1}{2} \sum_{i,k=1}^{K-1} L_{ik}^* \nabla \Gamma_i^* \cdot \nabla \Gamma_k^* \right] \mathrm{d}V + \oint_\Omega \Gamma_k^* \boldsymbol{J}_k \cdot \mathrm{d}\boldsymbol{\Omega} = \max. \tag{6.62}$$

Now by using the component balances

$$\varrho \dot{c}_k + \nabla \cdot \boldsymbol{J}_k = 0 \qquad (k = 1, 2, \ldots, K) \tag{6.63}$$

valid for our case and by requiring (just as it has been done in the case of the heat conduction) that $\boldsymbol{J}_k$ and—due to (6.63)—also $\varrho \dot{c}_k$ are fixed during the variation with respect to the parameters Γ_k^*, i.e. $\delta \boldsymbol{J}_k = 0$ and $\delta(\varrho \dot{c}_k) = 0$, furthermore that nothing is varied at the boundaries of the system, the extremum task (6.62) is already determined solely by the volume integral

$$\int_V \left[\sum_{k=1}^{K-1} \varrho \Gamma_k^* \dot{c}_k - \frac{1}{2} \sum_{i,k=1}^{K-1} L_{ik}^* \nabla \Gamma_i^* \cdot \nabla \Gamma_k^* \right] \mathrm{d}V = \max. \tag{6.64}$$

Here the integrand is nothing else than the Lagrangian density (in energy picture) of the diffusion problem:

$$\mathscr{L}^*_{\Gamma*} = \varrho \sum_{k=1}^{K-1} \Gamma^*_k \dot{c}_k - \frac{1}{2} \sum_{i,k=1}^{K-1} L^*_{ik} \nabla \Gamma^*_i \cdot \nabla \Gamma^*_k \tag{6.65}$$

where the varied parameter was also written as index. With this Lagrangian density, the Euler-Lagrange equations (6.54) lead to the differential equations

$$\varrho \dot{c}_k - \sum_{i=1}^{K-1} \nabla \cdot [L^*_{ik} \nabla (\mu_i - \mu_K)] = 0, \quad (k = 1, 2, \ldots, K-1) \tag{6.66}$$

if, according to (6.57) we change over from the parameters Γ^*_k to the chemical potentials. The latter equations are already essentially identical to the Fick equations of multi-component isothermal systems. This can easily be demonstrated. From the functions

$$\mu_k = \mu_k(\varrho_1, \varrho_2, \ldots, \varrho_{K-1}), \quad (k = 1, 2, \ldots, K-1) \tag{6.67}$$

which are valid in isothermal and isobar cases the identities

$$\nabla(\mu_i - \mu_K) = \sum_{j=1}^{K-1} \frac{\partial(\mu_i - \mu_K)}{\partial \varrho_j} \nabla \varrho_j, \quad (i = 1, 2, \ldots, K-1) \tag{6.68}$$

follow, with the use of which and, with the following expressions of the diffusion coefficients

$$\mathrm{D}^*_{kj} = \sum_{i=1}^{K-1} L^*_{ik} \frac{\partial(\mu_i - \mu_K)}{\partial \varrho_j}, \quad (k, j = 1, 2, \ldots, K-1) \tag{6.69}$$

from (6.66) the differential equations

$$\varrho \dot{c}_k = \sum_{j=1}^{K-1} \nabla \cdot (\mathrm{D}^*_{kj} \nabla \varrho_j), \quad (k = 1, 2, \ldots, K-1) \tag{6.70}$$

are obtained. If our diffusion system even satisfies the conditions that, the density ϱ is independent of time and it is uniform, the vanishing of the divergence of the barycentric velocity $\nabla \cdot \boldsymbol{v} = 0$, follows from the mass balance (2.17). Otherwise, if the boundary conditions of the diffusion system are such that the normal component of $\boldsymbol{v}$ vanishes at the walls of the system, $\boldsymbol{v}$ is zero everywhere in the system. This so-called "convection-free" case, which is a very particular one from the theoretical viewpoint occurs very often in practice. Instead of (6.70) (by taking also the transformation of the diffusion coefficients into account):

$$\frac{\partial c_k}{\partial t} = \sum_{j=1}^{K-1} \nabla \cdot (\mathrm{D}^*_{kj} \nabla c_j), \quad (k = 1, 2, \ldots, K-1) \tag{6.71}$$

can be given, which is the wellknown total set of the Fick type diffusion equations [3].

For the sake of the following, the Lagrangian density (6.65) will be transformed a bit. By writing the Gibbs-Duhem relation in our isothermal and isobar case in the form

$$\sum_{k=1}^{K-1} c_k \, \mathrm{d}\mu_k + \left(1 - \sum_{k=1}^{K-1} c_k\right) \mathrm{d}\mu_K = 0 \tag{6.72}$$

following from (3.28a), and then with (6.57) changing over to the parameters Γ_k^*, we get

$$\sum_{k=1}^{K-1} c_k \, \mathrm{d}\Gamma_k^* = \mathrm{d}\mu_K \tag{6.73}$$

both sides of which are total differentials. By using the relations

$$\mathrm{d}c_k = \sum_{j=1}^{K-1} \frac{\partial c_k}{\partial \Gamma_j^*} \mathrm{d}\Gamma_j^*, \quad (k = 1, 2, \ldots, K-1) \tag{6.74}$$

(following from the functions $c_k = c_k(\Gamma_1^*, \Gamma_2^*, \ldots, \Gamma_{k-1}^*)$ and the symmetry relations valid for the mixed second derivatives

$$g_{ik} \equiv \frac{\partial \mu_K}{\partial \Gamma_i^* \, \partial \Gamma_k^*} = \frac{\partial c_i}{\partial \Gamma_k^*} = \frac{\partial c_k}{\partial \Gamma_i^*} = \frac{\partial^2 \mu_K}{\partial \Gamma_k^* \, \partial \Gamma_i^*} \equiv g_{ki},$$
$$(i, k = 1, 2, \ldots, K-1) \tag{6.75}$$

it can be seen that the Lagrangian density (6.65) can be rewritten in general form:

$$\mathscr{L}_{\Gamma^*}^* = \varrho \sum_{i,k=1}^{K-1} g_{ik} \Gamma_i^* \dot{\Gamma}_k^* - \frac{1}{2} \sum_{i,k=1}^{K-1} L_{ik}^* \, \nabla \Gamma_i^* \cdot \nabla \Gamma_k^* . \tag{6.76}$$

According to the definition (6.57) of the parameters Γ_k^*, and to the variational prescription used by us, this Lagrangian density already depends on the chemical potentials and on their gradients only, i.e. it does not contain the time derivatives $\dot{\Gamma}_k^*$ as independent variables. With this Lagrangian density the Euler-Lagrange equations

$$\varrho \sum_{k=1}^{K-1} g_{ik} \dot{\Gamma}_k^* + \sum_{k=1}^{K-1} \nabla \cdot (L_{ik}^* \, \nabla \Gamma_k^*) = 0, \quad (i = 1, 2, \ldots, K-1) \tag{6.77}$$

are obtained, which are the general equations of multi-component diffusion. The transformation of these equations in terms of densities or in terms of arbitrary concentration (e.g. mole fractions) is very simple.

The foregoing can be generalized further in the direction that possible chemical reactions taking place between the components should also be taken into consideration. For this purpose, only the corresponding dissipation potential need be formed in the energy picture. This will be similar to the last two terms of the integrand of (5.66), hence

$$\Psi^* = \frac{1}{2} \sum_{i,k=1}^{K-1} L_{ik}^* \nabla \Gamma_i^* \cdot \nabla \Gamma_k^* + \frac{1}{2} \sum_{j,r=1}^{R} L_{jr}^* \sum_{i,k=1}^{K-1} \nu_{ij} \nu_{kr} \Gamma_i^* \Gamma_k^* \tag{6.78}$$

is the actual form of the dissipation potential by which—owing to (6.51) and (6.76)—we get the actual form of the Lagrangian density

$$\mathscr{L}_{\Gamma^*}^* = \varrho \sum_{i,k=1}^{K-1} g_{ik} \Gamma_i^* \dot{\Gamma}_k^* - \frac{1}{2} \sum_{i,k=1}^{K-1} L_{ik}^* \nabla \Gamma_i^* \cdot \nabla \Gamma_k^* - \frac{1}{2} \sum_{j,r=1}^{R} L_{jr}^* \sum_{i,k=1}^{K-1} \nu_{ij} \nu_{kr} \Gamma_i^* \Gamma_k^* . \tag{6.79}$$

By carrying out the differentiations indicated in (6.54), the Euler-Lagrange equations

$$\varrho \sum_{k=1}^{K-1} g_{ik} \dot{\Gamma}_k^* + \sum_{k=1}^{K-1} \nabla \cdot (L_{ik}^* \nabla \Gamma_k^*) = \sum_{j,r=1}^{R} L_{jr}^* \left(\sum_{k=1}^{K-1} \nu_{kr} \Gamma_k^* \right) \nu_{ij} = \sum_{j=1}^{R} \nu_{ij} J_j \tag{6.80}$$

are obtained. By these equations the evolution in space-time of the multi-component, isothermal, diffusing and reacting systems is described, i.e. the equations give an account of the component transport due to diffusion and, at the same time, also of the chemical reactions taking place in the system.

It should be noted that from the equations (6.80) the particular stationary equations, which are the special form of (5.69b), can also be obtained. Though it is the entropy picture which was used in Chapter V, whereas here the energy picture is applied, the difference—if we confine ourselves also in Chapter V, to the isothermal case—is irrelevant, since the parameters Γ_k and Γ_k^* defined in (5.65) and (6.57), respectively, only differ in cases of constant factor T. Therefore the Lagrangian densities of the corresponding entropy picture are T^{-1} times what was given hitherto, i.e.

$$\mathscr{L}_{\Gamma} = T^{-1} \mathscr{L}_{\Gamma^*}^* \tag{6.81}$$

is valid.

4. Deduction of the General Equation of Motion of Hydrodynamics

Now the general equation of motion of the viscous flow will be derived as a very specific example for the confirmation of the integral principle. The peculiarity of this example consists in the fact that in the case of viscous flow the thermodynamic force is determined by the velocity gradient $\nabla \boldsymbol{v}$, hence the variation must be carried out with respect to v_α ($\alpha = 1, 2, 3$) which is a typical "β" parameter as understood by Casimir.

The wellknown Navier-Stokes equation of the viscous flow was first deduced from the integral principle by VERHÁS [60, 74]. The deduction of the generalized Navier-Stokes equation is due to BÖRÖCZ, who deduced, by considering the antisymmetric part of the hydrodynamic pressure tensor, the more general form of the Navier-Stokes equation in which the rotational viscous term is included [75]. In the following, the most general form of the hydrodynamical equation of motion will be derived from the integral principle of thermodynamics. First of all, however, let us give the basic equations which are needed in order to give the variational condition with the aid of the actual form of $T\sigma_v$ and of Ψ_v^* in the case of viscous flow.

Since an isothermal case is examined, the force representation of the principle of least dissipation of energy can be given in the energy picture. From (3.68) in case of isothermal viscous fluids for the energy dissipation, we get

$$T\sigma_v = -p^v \, \nabla \cdot \boldsymbol{v} - \overset{0}{\mathbf{P}}{}^{vs} : \overset{0}{(\nabla \boldsymbol{v})}{}^s - \boldsymbol{P}^{va} \cdot (\nabla \times \boldsymbol{v} - 2\boldsymbol{\omega}) \geqq 0 \quad (6.82)$$

with the use of expressions (3.64), (3.65) and (3.66) of the forces. This expression can be transformed slightly, because the scalar product of the axial vectors $\boldsymbol{P}^{va}$ and $\nabla \times \boldsymbol{v}$ in the last term can also be given as the scalar product of the corresponding antisymmetric tensors, i.e.

$$-\boldsymbol{P}^{va} \cdot (\nabla \times \boldsymbol{v}) = \mathbf{P}^{va} : (\nabla \boldsymbol{v})^a = -\widetilde{\mathbf{P}}{}^{va} : (\nabla \boldsymbol{v})^a \quad (6.83)$$

where $\widetilde{\mathbf{P}}{}^{va} = -\mathbf{P}^{va}$ is the antisymmetric part of the transposed viscous pressure tensor $\widetilde{\mathbf{P}}{}^v$. With the help of the relations (6.83) and (2.103), the expression of the energy dissipation can also be given in the following concise form:

$$T\sigma_v = -\widetilde{\mathbf{P}}{}^v : \nabla \boldsymbol{v} + 2\boldsymbol{\omega} \cdot \boldsymbol{P}^{va} \geqq 0. \quad (6.84)$$

In (6.82) and (6.84) $\boldsymbol{\omega}$ denotes the mean angular velocity vector belonging to the internal rotation, which is determined in every point of the viscous fluid by the "rigid-like" rotation of the particles. For the kinetic energy of the internal rotation belonging to this angular velocity, the energy balance (2.164) is valid, which by considering (2.163b) can also be given in the form:

$$\varrho \Theta \boldsymbol{\omega} \cdot \dot{\boldsymbol{\omega}} = -2\boldsymbol{\omega} \cdot \boldsymbol{P}^{va}. \quad (6.85)$$

Now let us write Onsager's linear constitutive equations for the isotropic case. Considering a one-component viscous fluid (or by neglecting the chemical reaction in the multi-component case), in the case of isotropic continua the linear constitutive equations (3.91), (3.94) and

(3.95) can be written in the following explicit form:

$$p^{\mathrm{V}} = -\eta_{\mathrm{v}} \nabla \cdot \boldsymbol{v} = -\overset{(ss)}{L^*} \nabla \cdot \boldsymbol{v}$$

$$\boldsymbol{P}^{\mathrm{va}} = -\eta_{\mathrm{r}}(\nabla \times \boldsymbol{v} - 2\boldsymbol{\omega}) = -\overset{(aa)}{L^*}(\nabla \times \boldsymbol{v} - 2\boldsymbol{\omega}) \tag{6.86}$$

$$\overset{0}{\mathbf{P}}{}^{\mathrm{vs}} = -2\eta(\overset{0}{\nabla \boldsymbol{v}})^{\mathrm{s}} = -\overset{(tt)}{L^*}(\overset{0}{\nabla \boldsymbol{v}})^{\mathrm{s}}.$$

In these linear relations the coefficients $\overset{(ss)}{L^*}$, $\overset{(aa)}{L^*}$ and $\overset{(tt)}{L^*}$ are given in energy picture and are analogous to the coefficients given in (3.98), furthermore in the relation there are

$$\overset{(ss)}{L^*} \equiv \eta_{\mathrm{v}}, \quad \overset{(aa)}{L^*} \equiv \eta_{r}, \quad \overset{(tt)}{L^*} \equiv 2\eta \tag{6.87}$$

with the volume viscosity η_{v}, the shear viscosity η and the "rotational viscosity" η_{r}.

The form of the dissipation potential valid in our case can immediately be given with the linear constitutive equations (6.86):

$$\Psi^* \equiv \frac{\eta_{\mathrm{v}}}{2} (\nabla \cdot \boldsymbol{v})^2 + \eta \, [(\overset{0}{\nabla \boldsymbol{v}})^{\mathrm{s}} : (\overset{0}{\nabla \boldsymbol{v}})^{\mathrm{s}}] + \frac{\eta_{\mathrm{r}}}{2} (\nabla \times \boldsymbol{v} - 2\boldsymbol{\omega})^2, \tag{6.88}$$

where the ordinary dissipation function wellknown from hydrodynamics as Rayleigh function is determined by the first two terms. If we would confine ourselves to the more restricted task of deriving only the ordinary Navier-Stokes equation, then it would be sufficient to maintain the first two terms of (6.88) and to use the Rayleigh dissipation function only (see Verhás [60, 74]). However, it is worthwile to pay attention to the last term which is determined by the square of the force $\boldsymbol{X}_{\mathrm{v}}^{\mathrm{a}*} \equiv -(\nabla \times \boldsymbol{v} - 2\boldsymbol{\omega})$. (This force is determined by the vortices of the hydrodynamic velocity field and by the mean internal angular velocity of the constituent particles. Consequently, there exists an interaction between the macroscopic velocity field and the internal rotation of the particles only until $\nabla \times \boldsymbol{v} \neq 2\boldsymbol{\omega}$. Hence, an interaction related to an angular momentum transport of irreversible nature takes place between the macroscopic velocity field and the internal field of the particles as long as the last term of the dissipation potential (6.88) does not vanish. This interaction can be described with differential equations valid for $\boldsymbol{v}$ and $\boldsymbol{\omega}$ simultaneously. Our task is to derive these differential equations from the integral principle.

The dissipation potential (6.88) can also be given with the help of the relation (2.156) in the form

$$\Psi_{\mathrm{v}}^* \equiv \frac{1}{2}\left(\eta_{\mathrm{v}} - \frac{2}{3}\eta\right)(\nabla \cdot \boldsymbol{v})^2 + \eta \, [(\nabla \boldsymbol{v})^{\mathrm{s}} : (\nabla \boldsymbol{v})^{\mathrm{s}}] + \frac{\eta_{\mathrm{r}}}{2} [\nabla \times \boldsymbol{v} - 2\boldsymbol{\omega})^2. \tag{6.89}$$

With this expression of Ψ_{V}^{*} and with the form (6.84) of $T\sigma_{\mathrm{V}}$, the extremum principle

$$-\int\limits_V \left\{ \widetilde{\mathbf{P}}^{\mathrm{V}} : \nabla \boldsymbol{v} - 2\boldsymbol{\omega} \cdot \boldsymbol{P}^{\mathrm{va}} + \frac{1}{2}\left(\eta_{\mathrm{v}} - \frac{2}{3}\eta\right)(\nabla \cdot \boldsymbol{v})^2 + \eta[(\nabla \boldsymbol{v})^{\mathrm{s}} : (\nabla \boldsymbol{v})^{\mathrm{s}}] + \frac{\eta_{\mathrm{r}}}{2}(\nabla \times \boldsymbol{v} - 2\boldsymbol{\omega})^2 \right\} \mathrm{d}V = \max \tag{6.90}$$

is formulated in agreement with the (6.1) general variational condition. Of course, (6.90) is given in energy picture, on the other hand, by changing the sign the minimum could be written as it has been done by original authors [60, 74, 75]. In order to eliminate the first term of the integrand, we employ the kinetic energy balance

$$\varrho \frac{\mathrm{d}\frac{1}{2}\boldsymbol{v}^2}{\mathrm{d}t} = -\nabla \cdot (\boldsymbol{P} \cdot \boldsymbol{v}) + \widetilde{\mathbf{P}} : \nabla \boldsymbol{v} + \varrho \boldsymbol{F} \cdot \boldsymbol{v}, \tag{6.91}$$

which is equivalent to (2.141) and is valid for the case of an antisymmetric pressure tensor $\mathbf{P}$. This balance equation can easily be transformed by using the (2.109) decomposition of the pressure tensor, in the form

$$\boldsymbol{v} \cdot (\varrho \dot{\boldsymbol{v}} + \nabla p - \varrho \boldsymbol{F}) + \nabla \cdot (\mathbf{P}^{\mathrm{V}} \cdot \boldsymbol{v}) = \widetilde{\mathbf{P}}^{\mathrm{V}} : \nabla \boldsymbol{v}. \tag{6.92}$$

If we eliminate the first two terms of the integrand (6.90) with the balance equations (6.85) and (6.92), and the volume integral of the divergence term $\nabla \cdot (\mathbf{P}^{\mathrm{V}} \cdot \boldsymbol{v})$ is transformed to a surface integral by the Gauss theorem, the variational condition

$$-\delta \int\limits_V \left\{ \boldsymbol{v} \cdot (\varrho \dot{\boldsymbol{v}} + \nabla p - \varrho \boldsymbol{F}) + \varrho \Theta \boldsymbol{\omega} \cdot \dot{\boldsymbol{\omega}} + \frac{1}{2}\left(\eta_{\mathrm{v}} - \frac{2}{3}\eta\right)(\nabla \cdot \boldsymbol{v})^2 + \eta[(\nabla \boldsymbol{v})^{\mathrm{s}} : (\nabla \boldsymbol{v})^{\mathrm{s}}] + \frac{\eta_{\mathrm{r}}}{2}(\nabla \times \boldsymbol{v} - 2\boldsymbol{\omega})^2 \right\} \mathrm{d}V - \delta \oint\limits_{\Omega} (\mathbf{P}^{\mathrm{V}} \cdot \boldsymbol{v}) \cdot \mathrm{d}\boldsymbol{\Omega} = 0 \tag{6.93}$$

may be written instead of (6.90). With the necessary condition (6.93) of the maximum (6.90), the variation is now taken—just as always in the case of the integral principle derived from the force representation of the principle of least dissipation of energy—exclusively with respect to the internal variables of the forces. In the present example we have two such internal variables: $\boldsymbol{v}$ and $\boldsymbol{\omega}$. Hence with the variational task (6.93) we vary exclusively and independently with respect to $\boldsymbol{v}$ and $\boldsymbol{\omega}$ and all the other quantities are considered fixed during the variation. Hence, the time derivatives of the velocity $\dot{\boldsymbol{v}}$ and the angular velocity $\dot{\boldsymbol{\omega}}$ are not varied and, neither are varied p and $\boldsymbol{F}$ causing the non-dissipative processes,[1] finally nothing is varied along the boundary surface, i.e. $\mathbf{P}^{\mathrm{v}}$ and $\boldsymbol{v}$ are fixed at the boundaries. In case of the prescribed variational conditions, the volume integral of the necessary condition (6.93) determines the maximum prescribed in (6.90). Thus the Lagrangian density of the variational task is (in energy picture):

$$\mathscr{L}^*_{v\omega} = -\boldsymbol{v}\cdot(\varrho\dot{\boldsymbol{v}} + \nabla p - \varrho\boldsymbol{F}) - \varrho\Theta\boldsymbol{\omega}\cdot\dot{\boldsymbol{\omega}} - \frac{1}{2}\left(\eta_{\mathrm{v}} - \frac{2}{3}\eta\right)(\nabla\cdot\boldsymbol{v})^2$$

$$- \eta[(\nabla\boldsymbol{v})^{\mathrm{s}} : (\nabla\boldsymbol{v})^{\mathrm{s}}] - \frac{\eta_{\mathrm{r}}}{2}(\nabla\times\boldsymbol{v} - 2\boldsymbol{\omega})^2 \tag{6.94}$$

where the varied parameters were again written as indices.

According to the general formulation of the integral principle as given in (6.51) to (6.54), the following Euler-Lagrangian equations belong to the (6.94) Lagrangian density.

If $\Gamma_i \equiv -v_\beta$, we have

$$\frac{\partial\mathscr{L}^*_{v\omega}}{\partial v_\beta} - \sum_{\alpha=1}^{3}\frac{\partial}{\partial x_\alpha}\frac{\partial\mathscr{L}^*_{v\omega}}{\partial\left(\dfrac{\partial v_\beta}{\partial x_\alpha}\right)} = 0, \qquad (\beta = 1, 2, 3) \tag{6.95}$$

while, if $\Gamma_i \equiv -\omega_\beta$, we have

$$\frac{\partial\mathscr{L}^*_{v\omega}}{\partial \omega_\beta} - \sum_{\alpha=1}^{3}\frac{\partial}{\partial x_\alpha}\frac{\partial\mathscr{L}^*_{v\omega}}{\partial\left(\dfrac{\partial \omega_\beta}{\partial x_\alpha}\right)} = 0, \qquad (\beta = 1, 2, 3). \tag{6.96}$$

The actual forms of these simultaneously valid Euler-Lagrange equations can be determined by a simple but somewhat time-consuming calculation.

Let us consider primarily the first group of the Euler-Lagrange equations. By carrying out the differentiations in (6.95), we get

$$\varrho\dot{\boldsymbol{v}} + \nabla p - \varrho\boldsymbol{F} - \eta\,\Delta\boldsymbol{v} - \left(\frac{\eta}{3} + \eta_{\mathrm{v}}\right)\nabla\nabla\cdot\boldsymbol{v} - \eta_{\mathrm{r}}\,\nabla\times(2\boldsymbol{\omega} - \nabla\times\boldsymbol{v})$$

$$- 2(\nabla\boldsymbol{v})^{\mathrm{s}}\cdot\nabla\eta - (\nabla\cdot\boldsymbol{v})\nabla\left(\eta_{\mathrm{v}} - \frac{2}{3}\eta\right) + (2\boldsymbol{\omega} - \nabla\times\boldsymbol{v})\times\nabla\eta_{\mathrm{r}} = 0 \tag{6.97}$$

---

[1] More precise from the actual form

$$\varrho\dot{\boldsymbol{v}} + \nabla p + \nabla\cdot\overset{0}{\mathbf{P}}{}^{\mathrm{vs}} + \nabla\cdot\boldsymbol{P}^{\mathrm{va}} = \varrho\boldsymbol{F}$$

of the general impulse balance (2.71), by using the variational convention: $\delta\overset{0}{\mathbf{P}}{}^{\mathrm{vs}} = 0$ and $\delta\boldsymbol{P}^{\mathrm{va}} = 0$ for the irreversible parts of the total impulse current density $\mathbf{P}$, we have the following "transformed variational conventions" for the integral principle:

$$\delta(\varrho\dot{\boldsymbol{v}} + \nabla p - \varrho\boldsymbol{F}) = 0 \quad\text{and}\quad \delta(\varrho\Theta\dot{\boldsymbol{\omega}}) = 0$$

which, essentially are equivalent to the prescriptions mentioned above.

which is the general equation of motion in hydrodynamics. This can easily be seen. Indeed, with the assumption that the viscosity coefficients η, η and η_v are independent of the space coordinates, the equation

$$\varrho\dot{\boldsymbol{v}} + \nabla p - \varrho \boldsymbol{F} - \eta\, \Delta \boldsymbol{v} - \left(\frac{\eta}{3} + \eta_v\right) \nabla\nabla \cdot \boldsymbol{v} - \eta_r \nabla \times (2\boldsymbol{\omega} - \nabla \times \boldsymbol{v}) = 0 \tag{6.98}$$

is obtained, which is the generalized Navier-Stokes equation including the term of rotational viscosity. Of course, assuming that the coefficients η, η_v and η_r are constant in space (6.98) is obtained directly from (6.95) [75]. Nevertheless, the integral principle also now as in all cases hitherto, could give allowance for the dependence of the coefficients on the space coordinates.

Now we will be concerned with the second group (6.96) of the Euler-Lagrange equations. First of all it should be observed that the derivations of the angular velocity $\boldsymbol{\omega}$ with respect to the coordinates are not included in the (6.94) Lagrangian density. Therefore, the Euler-Lagrange equations (6.96) are reduced to the following simple form

$$\frac{\partial \mathscr{L}^*_{v\omega}}{\partial \omega_\beta} = 0, \quad (\beta = 1, 2, 3) \tag{6.99}$$

from which the differential equation

$$\varrho\Theta\dot{\boldsymbol{\omega}} - 2\eta_r(\nabla \times \boldsymbol{v} - 2\boldsymbol{\omega}) = 0 \tag{6.100}$$

is obtained, by which the evolution of $\boldsymbol{\omega}$ in the course of time is given. For example, in such a case when the vortex vector $\nabla \times \boldsymbol{v}$ is approximately uniform and constant, furthermore when it was zero at the beginning, (6.100) yields for the time variation of $\boldsymbol{\omega}$ the following solution

$$\boldsymbol{\omega}(t) = \frac{1}{2} \nabla \times \boldsymbol{v} \left(1 - e^{-\frac{t}{\tau}}\right) \tag{6.101}$$

where the relaxation time τ is determined by the relation

$$\tau = \frac{\varrho\Theta}{4\eta_r}. \tag{6.102}$$

It can be read off from the solution (6.101) that after a time the exponential term will vanish and the hydrodynamic vortex vector will become equal to double the angular velocity of the internal rotation. In this case, the force $\boldsymbol{X}^a_v \equiv -(\nabla \times \boldsymbol{v} - 2\boldsymbol{\omega})$ occuring in the linear constitutive equations (6.86) as well as the corresponding impulse current density will vanish, i.e. there is no more interaction between the vortices of the macroscopic velocity field and the internal rotational motion of the particles [See also what has been said in connection with

(2.169)]. This case corresponds to the case of a symmetric pressure tensor, to which the Lagrangian density belongs:

$$\mathcal{L}_v^* = -\boldsymbol{v}\cdot(\varrho\dot{\boldsymbol{v}} + \nabla p - \varrho\boldsymbol{F}) - \frac{1}{2}\left(\eta_v - \frac{2}{3}\eta\right)(\nabla\cdot\boldsymbol{v})^2$$
$$- \eta[(\nabla\boldsymbol{v})^s:(\nabla\boldsymbol{v})^s]. \tag{6.103}$$

With this Lagrangian density the Euler-Lagrange equations (6.95) lead directly to the ordinary Navier-Stokes equation

$$\varrho\dot{\boldsymbol{v}} + \nabla p - \varrho\boldsymbol{F} - \eta\,\Delta\boldsymbol{v} - \left(\frac{\eta}{3} + \eta_v\right)\nabla\nabla\cdot\boldsymbol{v} = 0 \tag{6.104}$$

which follows, of course, from (6.98) if $\nabla\times\boldsymbol{v} = 2\boldsymbol{\omega}$ is valid in every time.

It is proved by the presented equations of hydrodynamical motion that with the aid of the integral principle the pure mechanical (inertial) motion and the dissipative effects superposed on it can be treated simultaneously. It is evident that the Reynolds equations of turbulent flow can also be derived from the integral principle, and the situation is similar with respect to the fundamental equations of magnetohydrodynamics, plasma physics, etc.

5. Non-isothermal Transport Equations

Up to now, always an indirect and inductive way was followed in the derivation of the transport equations. Essence of this procedure was that starting from the force representation of the principle of least dissipation of energy, in actual cases the Lagrangian density of the problem and the Euler-Lagrange equations were determined with the prescription of the way of variation and of the boundary conditions. In other words, in the case of the presented examples we have arrived at the integral principle in an inductive way, i.e. its existence was confirmed in an indirect manner. Now rather the reverse way will be followed, and the differential equations are governing non-isothermal transport processes (thermodiffusion, etc.) will be derived from the forms (6.51) to (6.54) of the generally formulated integral principle in a direct way [76]. It is verified by this method that we can consider the integral principle as an independent variational principle of thermodynamics.

The Lagrangian density valid in our case can be immediately given on the basis of the general form (6.51a) of it. Indeed, with the help of (3.19) expressing the substantial time derivatives of the Gibbs relation, we have

$$\varrho\dot{s} = \varrho\left(\frac{\dot{u}}{T} + \frac{p}{T}\dot{v} - \sum_{k=1}^{K}\frac{\mu_k}{T}\dot{c}_k\right), \tag{6.105}$$

by which the first "kinetic" term of the Lagrangian density is already determined. In our case the entropy production is determined by the second and third terms of (3.69), i.e.

$$\sigma_{\mathrm{qd}} = \boldsymbol{J}_{\mathrm{q}} \cdot \nabla\left(\frac{1}{T}\right) + \sum_{k=1}^{K} \boldsymbol{J}_k \cdot \left[\frac{\boldsymbol{F}_k}{T} - \nabla\left(\frac{\mu_k}{T}\right)\right], \tag{6.106}$$

where the expressions (3.71) and (3.72) of the forces were used. For simplicity's sake in (6.106), and further on also, we work with a dependent set of the diffusional fluxes $\boldsymbol{J}_k$, and so now their number will not be reduced to a system containing $K - 1$ independent fluxes. The following linear constitutive equations

$$\boldsymbol{J}_{\mathrm{q}} = L_{\mathrm{qq}} \nabla\left(\frac{1}{T}\right) + \sum_{k=1}^{K} L_{\mathrm{q}k} \left[\frac{\boldsymbol{F}_k}{T} - \nabla\left(\frac{\mu_k}{T}\right)\right], \tag{6.107}$$

$$\boldsymbol{J}_i = L_{i\mathrm{q}} \nabla\left(\frac{1}{T}\right) + \sum_{k=1}^{K} L_{ik} \left[\frac{\boldsymbol{F}_k}{T} - \nabla\left(\frac{\mu_k}{T}\right)\right], \quad (i = 1, 2, \ldots, K) \tag{6.108}$$

and reciprocal relations

$$L_{i\mathrm{q}} = L_{\mathrm{q}i}, \qquad (i = 1, 2, \ldots, K) \tag{6.109}$$

$$L_{ik} = L_{ki}, \qquad (i, k = 1, 2, \ldots, K) \tag{6.110}$$

are valid in the present case, since according to the theorem mentioned in Chapter IV sec. 4. a homogeneous linear dependency amongst the fluxes (for example, such as (1.43) is valid for the fluxes of the diffusion) leaves the validity of the reciprocal relations unimpaired. If we choose the Γ_i parameters in the general form of the integral principle (6.53) in such a manner that

$$\Gamma_{\mathrm{q}} \equiv \frac{1}{T}, \qquad \Gamma_k \equiv -\frac{\mu_k}{T}, \qquad (k = 1, 2, \ldots, K) \tag{6.111}$$

the forces can be given in the form

$$\boldsymbol{X}_{\mathrm{q}} \equiv \nabla \Gamma_{\mathrm{q}}, \quad \boldsymbol{X}_k \equiv \Gamma_{\mathrm{q}} \boldsymbol{F}_k + \nabla \Gamma_k, \quad (k = 1, 2, \ldots, K) \tag{6.112}$$

according to which the local dissipation potential is:

$$\Psi_{\mathrm{qd}} \equiv \frac{1}{2} L_{\mathrm{qq}} (\nabla \Gamma_{\mathrm{q}})^2 + \sum_{k=1}^{K} L_{\mathrm{q}k} \nabla \Gamma_{\mathrm{q}} \cdot (\Gamma_{\mathrm{q}} \boldsymbol{F}_k + \nabla \Gamma_k)$$
$$+ \frac{1}{2} \sum_{i,k=1}^{K} L_{ik} (\Gamma_{\mathrm{q}} \boldsymbol{F}_i + \nabla \Gamma_i) \cdot (\Gamma_{\mathrm{q}} \boldsymbol{F}_k + \nabla \Gamma_k). \tag{6.113}$$

With this dissipation potential and with (6.105) the Lagrangian density of the present problem is already determined. Indeed, the choice (6.111) of the parameters Γ_i taken into account also in (6.105), we can write

$$\mathscr{L} = \varrho\left(\Gamma_q \dot{u} + \Gamma_q p \dot{v} + \sum_{k=1}^{K} \Gamma_k \dot{c}_k\right)$$
$$- \left[\frac{1}{2} L_{qq}(\nabla \Gamma_q)^2 + \sum_{k=1}^{K} L_{qk} \nabla \Gamma_q \cdot (\Gamma_q \boldsymbol{F}_k + \nabla \Gamma_k) \right. \tag{6.114}$$
$$\left. + \frac{1}{2} \sum_{i,k=1}^{K} L_{ik}(\Gamma_q \boldsymbol{F}_i + \nabla \Gamma_i) \cdot (\Gamma_q \boldsymbol{F}_k + \nabla \Gamma_k)\right]$$

which is the Lagrangian density of our problem in entropy picture.

In the possession of the Lagrangian density our task is only to carry out the differentiation denoted in the general form of the Euler-Lagrange equations (6.54) with respect to the parameters Γ_q and Γ_k of (6.111). In the outlined way, the transport equation of energy

$$\varrho \dot{u} + \varrho p \dot{v} - \sum_{k=1}^{K} L_{qk} \nabla \Gamma_q \cdot \boldsymbol{F}_k - \sum_{i,k=1}^{K} L_{ik}(\Gamma_q \boldsymbol{F}_i + \nabla \Gamma_i) \cdot \boldsymbol{F}_k$$
$$+ \nabla \cdot \left[L_{qq} \nabla \Gamma_q + \sum_{k=1}^{K} L_{qk}(\Gamma_q \boldsymbol{F}_k + \nabla \Gamma_k)\right] = 0 \tag{6.115}$$

and the equations governing the components transport

$$\varrho \dot{c}_i + \nabla \cdot \left[L_{qi} \nabla \Gamma_q + \sum_{k=1}^{K} L_{ik}(\Gamma_q \boldsymbol{F}_k + \nabla \Gamma_k)\right] = 0, \quad (i = 1, 2, \ldots, K) \tag{6.116}$$

are obtained.

The above deduced-transport equations are general ones, because they give an account of internal energy and components transport occurring in non-isothermal and multi-component continua under the influences of arbitrary external force. By restricting ourselves in the following to an incompressible ($\dot{v} = 0$) fluid, where no external forces are present ($\boldsymbol{F}_k = 0$), (6.115) and (6.116) yield the wellknown transport equations of thermodiffusional systems

$$\varrho \dot{u} = \nabla \cdot \left[-L_{qq} \nabla\left(\frac{1}{T}\right) + \sum_{k=1}^{K} L_{qk} \nabla\left(\frac{\mu_k}{T}\right)\right], \tag{6.117}$$

and

$$\varrho \dot{c}_i = \nabla \cdot \left[-L_{qi} \nabla\left(\frac{1}{T}\right) + \sum_{k=1}^{K} L_{ik} \nabla\left(\frac{\mu_k}{T}\right)\right], \quad (i = 1, 2, \ldots, K) \tag{6.118}$$

meanwhile the relations (6.111) have also been used.

We want to note that a generalization of the above transport equations by which chemical reactions could be described is simple. In this case only the chemical terms corresponding to the second sum of the dissipation potential given in (6.78)—but, of course, expressed in term of the entropy picture now—must be added to the Lagrangian density (6.114).

6. Deduction of the Transport Equations in a General Form

Hitherto the most important basic equations of multi-component, hydro-thermodynamic systems were derived and it was found that the integral principle determined by the relations (6.51) to (6.54) have general validity. Up to now, however, we omitted the derivation of a general form of the transport equations from the integral principle, and this will be given in the following.

Let us start from the global form of the principle of least dissipation of energy written in the force representation

$$\delta \int_V [\varrho \dot{s} + \nabla \cdot \boldsymbol{J}_s - \Psi] \, \mathrm{d}V = 0 \tag{6.119}$$

from which, applying the Gauss divergence theorem, we have

$$\delta \int_V [\varrho \dot{s} - \Psi] \, \mathrm{d}V + \delta \oint_\Omega \boldsymbol{J}_s \cdot \mathrm{d}\boldsymbol{\Omega} = 0 . \tag{6.120}$$

Let us now consider the specific entropy s quite generally as expressed in terms of the specific values a_i of the f independent extensive A_i quantity:

$$s = s(a_1, a_2, \ldots, a_f) . \tag{6.121}$$

From this follows

$$\dot{s} = \sum_{i=1}^{f} \frac{\partial s}{\partial a_i} \dot{a}_i = \sum_{i=1}^{f} \Gamma_i \dot{a}_i , \tag{6.122}$$

which is the substantial time derivative of the generalized Gibbs relation already discussed in (3.25). Assuming that the dissipation potential Ψ includes only forces which are the gradients of the parameters Γ_i (this means that we confine ourselves to transport processes only, but this does not represent a considerable restriction of generality, because e.g. the chemical reactions could always be accounted for in a way similar to (6.78)) then instead of (6.120) the following equation may be written

$$\delta \int_V \left[\varrho \sum_{i=1}^{f} \Gamma_i \dot{a}_i - \frac{1}{2} \sum_{i,k=1}^{f} L_{ik} \nabla \Gamma_i \cdot \nabla \Gamma_k \right] \mathrm{d}V = 0 \tag{6.123}$$

with the condition that the varied quantities vanish along the boundaries. The condition (6.123) is considered as such a variational task by which the variation was carried out solely with respect to the internal variables of the forces, i.e. over the parameters Γ_i. Proceeding in this way, we have from (6.123)

$$\int_V \left[\varrho \dot{a}_i + \sum_{k=1}^{f} \nabla \cdot (L_{ik}\, \nabla \Gamma_k)\right] \delta \Gamma_i \,\mathrm{d}V = 0 \tag{6.124}$$

which is fulfilled in case of an arbitrary variation $\delta \Gamma_i$ only if

$$\varrho \dot{a}_i = - \sum_{k=1}^{f} \nabla \cdot (L_{ik}\, \nabla \Gamma_k), \quad (i = 1, 2, \ldots, f). \tag{6.125}$$

These equations are general forms of the transport equations and, at the same time, they are the Euler-Lagrange equations belonging to the integral principle (6.123) in agreement with (6.54).

In the above, the general forms of the transport equations determined by the integral principle (6.53) were deduced in a general way. However, it is advisable to transform the Lagrangian density

$$\mathscr{L} = \varrho \dot{s} - \Psi = \varrho \sum_{i=1}^{f} \Gamma_i \dot{a}_i - \frac{1}{2} \sum_{i,k=1}^{f} L_{ik}\, \nabla \Gamma_i \cdot \nabla \Gamma_k. \tag{6.126}$$

For this purpose, let us start from the set of functions

$$a_i = a_i(\Gamma_1, \Gamma_2, \ldots, \Gamma_f), \; (i = 1, 2, \ldots, f) \tag{6.127}$$

from which follows

$$\dot{a}_i = \sum_{k=1}^{f} \frac{\partial a_i}{\partial \Gamma_k} \dot{\Gamma}_k = \sum_{k=1}^{f} s_{ik}^{-1}\, \dot{\Gamma}_k, \; (i = 1, 2, \ldots, f) \tag{6.128}$$

where s_{ik}^{-1} is the reciprocal of the matrix s_{ik} determined by the second derivatives of the entropy function (6.121) and for the elements of which the Maxwellian reciprocal relations

$$s_{ik}^{-1} \equiv \frac{\partial^2 s}{\partial \Gamma_i \partial \Gamma_k} = \frac{\partial a_i}{\partial \Gamma_k} = \frac{\partial a_k}{\partial \Gamma_i} = \frac{\partial^2 s}{\partial \Gamma_k \partial \Gamma_i} \equiv s_{ki}^{-1}, \quad (i, k = 1, 2, \ldots, f) \tag{6.129}$$

are valid. Evidently, the expressions (6.75) already used in the case of diffusion are the particular forms of (6.129). The transformed form of the Lagrangian density is readily obtained from (6.126) with (6.128):

$$\mathscr{L} = \varrho \sum_{i,k=1}^{f} s_{ik}^{-1} \Gamma_i \dot{\Gamma}_k - \frac{1}{2} \sum_{i,k=1}^{f} L_{ik} \nabla \Gamma_i \cdot \nabla \Gamma_k. \tag{6.130}$$

This form of the Lagrangian density differs from that given in (6.126) only in so far that instead of the time derivative of the specific values $\dot{a}_i$ it is expressed in terms of time derivatives of the Γ_k parameters. By starting from the Lagrangian density (6.130), the Euler-Lagrange equations (6.54) belonging to the integral principle are obtained in the form

$$\varrho \sum_{k=1}^{f} s_{ik}^{-1} \dot{\Gamma}_k + \sum_{k=1}^{f} \nabla \cdot (L_{ik} \nabla \Gamma_k) = 0, \quad (i = 1, 2, \ldots, f) \quad (6.131)$$

which are, of course, equivalent to (6.125) in every respect.

With the above deduction of the general form of the transport equations we started from the formulation of the principle of least dissipation of energy given on the left-hand side of the force representation (6.1), we reached our aim directly with the aid of the substantial time derivative (6.122) of the generalized Gibbs relation. This method, which can be regarded as a direct one, was applied in case of the deduction of the equation of heat conduction in Fourier's pictures and in entropy picture too, furthermore, also in the derivation of the non-isothermal transport equations.

As a matter of course, the derivation of the general transport equations can also be done by starting from the right-hand side of equation (6.1). In this case, however, the divergencies of the different current-densities occurring in the entropy production can be eliminated only with the corresponding balance equations. More exactly, the time derivatives $\dot{a}_i$ can be introduced into the Lagrangian density only in an indirect way, and this is the reason that this method can be called an indirect one. The method has first been used by Verhás [60, 74], and in particular his way of though was followed by us in the derivation of the equation of heat conduction in energy picture and in the case of the generalized "Γ" picture, further, in the derivation of the general equation of motion of the viscous flow and in the case of Fick's equation of the isothermal diffusion as well. Since the essential part of this indirect method can easily be seen from the special cases (particularly from the method followed in the derivation of the Fourier equation in case of the generalized "Γ" picture), therefore the derivation of the general form of transport equations is disregarded here.

It can easily be seen that the integral principle formulated in equations (6.51) to (6.54) has an intrinsic relation to the principle of the least dissipation of energy, more precise, with its formulation given in force representation. It can be stated that the integral principle is equivalent to the existence of the differential equations governing the evolution in space-time of the Γ_i parameters, which are the internal variables of the thermodynamic forces. Therefore, though the integral principle determined by the relations (6.51) to (6.54) is also self-consistent, its close relation to the principle of least dissipation of energy formulated in force representation is doubtless. This was also indicated by the prescription of the variation, since in the case of the force representation of the principle of least dissipation of energy we vary only with respect to the forces considered as the causes of the irreversible processes, whereas in case of the integral principle we vary with respect to the internal variables Γ_i. Summaryzing: the integral principle is valid in the case where the variation has been prescribed by the following conventions:

1. *The variation is carried out solely with respect to those parameters Γ_i, which are the internal variables of the forces considered as the causes of the irreversible processes. Particularly in the case of transport processes the $\boldsymbol{X}_i$ are the gradients of Γ_i.*

2. *Nothing is varied along the boundary surfaces of the continuum considered, i.e. each current density and also every other parameter there is considered as prescribed with respect to the variation.*[1]

The validity of the integral principle is unambiguously ensured by these conditions of the variational conventions; however the above conditions should be complemented by some remarks.

The integral principle is directly equivalent to the existence of the total set of differential equations governing dissipative processes. However, it also includes in a particular and implicit way the equations valid for reversible motions. For example, in the case of the equation of motion of hydrodynamics the terms, by which the pure mechanical reversible motion (without viscosity and dissipation) is described, are also included in the equation. This is a consequence of the fact that the external forces causing reversible motions are considered as prescribed during the variation. The case is similar to the non-variation of the time derivatives $\dot{a}_i$ or $\dot{\Gamma}_i$, which is evidently due to the fluxes being fixed and therefore a consequence of the fundamental prescription of the force representation. We will return to this problem in quite a general way in the following section, when the relation of the integral principle to the Hamilton principle will be analysed.

In the possession of the results it is emphasized again that the force representation of the principle of least dissipation of energy is in each respect, i.e. in stationary as well as in general cases more productive than the flux representation. The differential equations governing irreversible processes may be determined, at least in a direct way, only from the force representation. However, it should be noted that if we assume the existence of general "velocity potentials" ξ_i by the gradients of which the current densities are determined, i.e.

$$\boldsymbol{J}_i = \nabla \xi_i$$

[1] The validity of the integral principle can also be extended for the case of "free boundary conditions" by removing the mentioned constraints. This fact, for example, can readily be verified from formulation of (6.120) which also includes the surface term.

the differential equations formally similar to the wellknown transport equations can be obtained from the flux representation as well. Such "velocity potentials," however, have no direct physical meaning, whereas the parameters Γ_i do have, moreover these and not the forces $X_i \equiv \nabla \Gamma_i$, are the quantities which can be measured in a direct, experimental way. The latter fact also means that the integral principle formulated in (6.51) to (6.54) is from the practical viewpoint also the most useful principle of non-equilibrium thermodynamics, since, it is valid for those intensive equilibrium state parameters which can experimentally be measured directly (temperature, chemical potentials or concentrations, velocities, etc.).

With a view to recent results of VOJTA, the foregoing requires some completion [77, 79]. It has been demonstrated by VOJTA that Onsager's original formulation of the principle of least dissipation of energy in flux representation, i.e. (4.86) or rather (4.90) can also be given for the case of continua with the aid of the functional formalism. In the Vojta approach for the global OM functional an integral principle similar to that given in (4.90), is valid. Consequently in the case of the functional formulation an integral principle exists for dissipative continua, which refers to the extremum of a time integral. Vojta's approach is relevant not only because it gives the general formulation of the principle of least dissipation of energy in the flux representation, but also because by this development the statistical meaning of the variational principle is made possible according by the applicability of the fluctuation theory [77].

7. Relations between Integral Principle and Hamilton Principle

It is certain that the integral principle generally formulated in equations (6.51) to (6.54), shows some similarity to the integral principles used in other branches of physics, particularly to the Hamilton principle. It is our task now to examine in detail the similarities and differences between the integral principle of thermodynamics and the Hamilton principle. Though it will be seen that this analysis does not lead to essentially new results, it is very useful, notably, with its aid the relations between the reversible and dissipative processes can be unambiguously determined. Simualtaneously, we also get an answer in a quite general manner because in the case of the integral principle of thermodynamics the time derivatives of the state parameters must not be varied.

First of all it should be pointed out that according to the general and explicit expression (6.130), the thermodynamic Lagrangian density $\mathscr{L}$ is a function of the parameters Γ_i and of their gradients $\nabla \Gamma_i$.

Hence

$$\mathscr{L} = \mathscr{L}(\Gamma_i, \nabla\Gamma_i) = \varrho\dot{s}(\Gamma_i, \dot{\Gamma}_i) - \Psi(\nabla\Gamma_i, \nabla\Gamma_i), \qquad (6.132)$$

since in the thermodynamic Lagrangian density the time derivatives are not new independent variables due to the balance equations and variational prescription.[1] Otherwise in (6.132) all $\Gamma_i = \Gamma_i(\boldsymbol{r}, t)$ quantities represent real and scalar field quantities. In this sense we may speak of a temperature field of a material continuum, of its chemical potential fields, and its velocity field (which is, of course, equivalent to three scalar fields). Accordingly, the (6.54) Euler-Lagrange equations always refer to real and scalar macroscopic field quantities dependent on space and time. In the following, the relation of the integral principle of thermodynamics to the Hamilton principle will be analysed based upon this fact and with the aid of the mathematical apparatus of the scalar field theories.

In order to examine the relation between the integral principle and the field theoretical form of the Hamilton principle it should be taken into consideration that the field quantities $\Gamma_i(\boldsymbol{r}, t)$—as generalized coordinates—form a continuum which is identical to an f dimensional abstract space, when the number of the independent parameters Γ_i is f. Therefore, let us interpret a Lagrangian density $\mathscr{L}$ of which it is assumed (similarly to the case of the Hamilton principle) that it depends on the "space coordinates" Γ_i, on its gradients $\nabla\Gamma_i$, and on its time derivatives $\dot{\Gamma}_i$, moreover, perhaps it also depends explicitly on time t. Hence, the Lagrangian density will be

$$\mathscr{L} = \mathscr{L}(\Gamma_i, \nabla\Gamma_i, \dot{\Gamma}_i, t). \qquad (6.133)$$

By integrating this over the volume of the continuum, we get for the (global) Lagrange function

$$L = \int_V \mathscr{L}(\Gamma_i, \nabla\Gamma_i, \dot{\Gamma}_i, t)\, \mathrm{d}V. \qquad (6.134)$$

Now the Hamilton principle is represented by the integral principle

$$\delta \int_{t_1}^{t_2} L\, \mathrm{d}t = \delta \int_{t_1}^{t_2} \int_V \mathscr{L}\, \mathrm{d}V\, \mathrm{d}t = \int_{t_1}^{t_2} \int_V (\delta\mathscr{L})\, \mathrm{d}V\, \mathrm{d}t = 0, \qquad (6.135)$$

if we agree that the field quantities Γ_i at the moments t_1 and t_2 are not varied, furthermore, that the time is not varied at all. Accordingly, the variational conventions are determined by the conditions

$$\delta\Gamma_i(\boldsymbol{r}_1, t_1) = \delta\Gamma_i(\boldsymbol{r}_1, t_2) = 0 \text{ and } \delta t = 0. \qquad (6.136)$$

[1] Of course, for example, in case of chemical reactions (and other scalar phenomena) Ψ depends themselves Γ_i parameters too. [See, e.g. (6.79)].

The conditions $\delta\Gamma_i(\mathbf{r}_1, t_1) = \delta\Gamma_i(\mathbf{r}_1, t_2) = 0$ mean that the position of the system is given for the time t_1 and t_2 and no variations are allowed at these limits. Therefore, we can say that by convention we vary between definite limits, because the limiting positions of the system are prescribed. With the expression (6.133) of the Lagrangian density, we get

$$\delta\mathscr{L} = \sum_{i=1}^{f}\left\{\frac{\partial\mathscr{L}}{\partial\Gamma_i}\,\delta\Gamma_i + \sum_{\alpha=1}^{3}\frac{\partial\mathscr{L}}{\partial\left(\frac{\partial\Gamma_i}{\partial x_\alpha}\right)}\,\delta\left(\frac{\partial\Gamma_i}{\partial x_\alpha}\right) + \frac{\partial\mathscr{L}}{\partial\dot{\Gamma}_i}\,\delta\dot{\Gamma}_i\right\} \qquad (6.137)$$

where in the last two terms the commutability of the variation (δ-process) and the differentiation (d- or ∂-process) was employed. By substituting expression (6.137) into (6.135) and after partial integration with conditions (6.136), we have

$$\int_{t_1}^{t_2}\int_V \sum_{i=1}^{f}\left\{\frac{\partial\mathscr{L}}{\partial\Gamma_i} - \sum_{\alpha=1}^{3}\frac{\partial}{\partial x_\alpha}\,\frac{\partial\mathscr{L}}{\partial\left(\frac{\partial\Gamma_i}{\partial x_\alpha}\right)} - \frac{\mathrm{d}}{\mathrm{d}t}\left(\frac{\partial\mathscr{L}}{\partial\dot{\Gamma}_i}\right)\right\}\delta\Gamma_i\,\mathrm{d}V\,\mathrm{d}t = 0 \qquad (6.138)$$

from which in case of arbitrary variations $\delta\Gamma_i$ it follows that

$$\frac{\partial\mathscr{L}}{\partial\Gamma_i} - \sum_{\alpha=1}^{3}\frac{\partial}{\partial x_\alpha}\,\frac{\partial\mathscr{L}}{\partial\left(\frac{\partial\Gamma_i}{\partial x_\alpha}\right)} - \frac{\mathrm{d}}{\mathrm{d}t}\left(\frac{\partial\mathscr{L}}{\partial\dot{\Gamma}_i}\right) = 0, \quad (i = 1, 2, \ldots, f). \qquad (6.139)$$

These differential equations are the Euler-Lagrange equations of the variational task (6.135) valid for the Lagrangian density (6.133), and these govern the evolution of the $\Gamma_i = \Gamma_i(\mathbf{r}, t)$ field quantities in space-time.

By the variational condition (6.135) together with the prescriptions (6.136) referring to the variational conventions, the Hamilton principle is represented. In other words, the way of variation, which is arbitrary in the case of each integral principle to a certain—but well-defined—extent by convention, is prescribed in the Hamilton principle by the conditions (6.136). It is wellknown, however, that with variational prescriptions more general as (6.136) other integral principles, which are different from the Hamilton principle, are obtained. Thus, for example, the Maupertuis principle, the principle of least action can be attained by the convention that the time is also varied (see [43, 44] and particularly [58]). In this sense, the Hamilton principle is a more restricted one than the principle of least action, since in the latter case the ensemble of the concurrent functions is larger. Comparing now the expressions (6.51) to (6.54) of the thermodynamic integral principle (with the general form (6.132) of the Lagrangian density) to the expressions (6.133) to (6.136) and (6.139), representing the Hamilton principle, the similarities and differences of the two principles become evident. These can be summarized as follows:

1. *The Lagrangian density of thermodynamics does not explicitly depend on time, whereas in the case of the Hamilton principle there is no restriction in this respect.*

2. *The integral principle of thermodynamics refers to the stationary (extremum) value of a volume integral, whereas the Hamilton principle refers to a definite time-integral.*

3. *The variations with respect to the field quantities Γ_i must be taken at a fixed time, in the case of both principles, i.e. time is not varied in either cases.*

4. *The time derivatives $\dot{\Gamma}_i$ of the field quantities Γ_i are varied in the case of the Hamilton principle, whereas they are not varied in the case of the integral principle of thermodynamics.*

The consequence of this last and very important difference between the two principles consists in the variational convention and occurs in the term

$$\frac{\mathrm{d}}{\mathrm{d}t}\left(\frac{\partial \mathscr{L}}{\partial \dot{\Gamma}_i}\right), \tag{6.140}$$

which is read off from the Euler-Lagrange equations (6.54) and (6.139) belonging to the two principles. This term evidently does not appear in thermodynamics, because this is the time derivative of the impulse density by which the Newtonian force density is determined. However, the irreversible transport processes (and every other irreversible process as well) are not the result of Newtonian forces, but of dissipative forces, which are the gradients (or linear combinations as in the case of chemical reactions) of the Γ_i parameters. On the basis of the foregoing it is evident that the different variational conventions with respect to the case of the two principles refer to the fundamental difference which exists between the reversible mechanical motion and the dissipative processes.

The difference between the thermodynamical integral principle and the Hamiltonian principle can be demonstrated more profoundly by recalling the formulation of the latter used in point mechanics. In the action integral of point mechanics

$$\delta \int_{t_1}^{t_2} L \,\mathrm{d}t = 0 \tag{6.141}$$

the Lagrange function

$$L = L(q_1, q_2, \ldots, q_f, \dot{q}_1, \dot{q}_2, \ldots, \dot{q}_f, t) \tag{6.142}$$

is a function of the general coordinates $q_1, q_2, \ldots, q_f$, and of the general velocities $\dot{q}_1, \dot{q}_2, \ldots, \dot{q}_f$ and occasionally of the time t. As is wellknown, the Lagrange equations of motion

$$\frac{\partial L}{\partial q_i} - \frac{\mathrm{d}}{\mathrm{d}t}\frac{\partial L}{\partial \dot{q}_i} = 0, \qquad (i = 1, 2, \ldots, f) \tag{6.143}$$

belong to the variational problem (6.141).

Let us come back to the form (6.135) of the Hamilton principle to which the Euler-Lagrange equations (6.139) belong. The forms of these field equations are different from those of the Lagrange equations (6.143) which are valid for the mass points. In order to bring the field equations, valid in the case of continua, to a form similar to the Lagrange equations of point mechanics, we have to change from the Lagrangian density $\mathscr{L}$ in the Euler-Lagrange equations (6.139) to the global $L = \int_V \mathscr{L}\,\mathrm{d}V$ Lagrange function. The transformation can be carried out by the following functional (variational) derivatives [44]:

$$\frac{\delta L}{\delta \dot{\Gamma}_i} = \frac{\partial \mathscr{L}}{\partial \dot{\Gamma}_i}, \qquad (i = 1, 2, \ldots, f) \tag{6.144}$$

and

$$\frac{\delta L}{\delta \Gamma_i} = \frac{\partial \mathscr{L}}{\partial \Gamma_i} - \sum_{\alpha=1}^{3} \frac{\partial}{\partial x_\alpha} \frac{\partial \mathscr{L}}{\partial\left(\dfrac{\partial \Gamma_i}{\partial x_\alpha}\right)}, \qquad (i = 1, 2, \ldots, f). \tag{6.145}$$

With these derivatives the field equations (6.139) can be given in the form

$$\frac{\delta L}{\delta \Gamma_i} - \frac{\mathrm{d}}{\mathrm{d}t}\frac{\delta L}{\delta \dot{\Gamma}_i} = 0, \qquad (i = 1, 2, \ldots, f) \tag{6.146}$$

which are already similar to the Lagrange equations (6.143) of particle mechanics. It is evident that with the formulation of the principles of point mechanics in the case of field theory the fundamental rôle is due to the functional derivatives of the type (6.144) and (6.145). Therefore, to understand the relation between the thermodynamic integral principle and Hamilton principle in the case of continua, an exceptional attention must be directed to these functional derivatives.

The functional derivatives in question play a very particular but briefly clarifiable rôle in thermodynamics. The derivatives (6.144) do not occur in the variational tasks of thermodynamics, according to the variational convention $\delta \dot{\Gamma}_i = 0$. On the other hand, the comparison of the general form (6.54) of thermodynamical Euler-Lagrange equations to (6.145) shows that

$$\frac{\delta L}{\delta \Gamma_i} = \frac{\partial \mathscr{L}}{\partial \Gamma_i} - \sum_{\alpha=1}^{3} \frac{\partial}{\partial x_\alpha} \frac{\partial \mathscr{L}}{\partial\left(\dfrac{\partial \Gamma_i}{\partial x_\alpha}\right)} = 0, \qquad (i = 1, 2, \ldots, f) \tag{6.147}$$

is valid. Hence the vanishing of these functional derivatives is equivalent to the existence of the Euler-Lagrange equations, which are identical to the transport equations. Thus the time-extremisation of with thermodyna mical variational tasks of the dissipation integral

$$L = \int_V \mathscr{L} \, \mathrm{d}V = \int_V (\varrho \dot{s} - \Psi) \, \mathrm{d}V \tag{6.148}$$

— similar to the time-extremisation of the action integral (6.135) — is immaterial. This is the reason why the functional derivatives (6.144) and (6.145) belonging to the Hamilton principle of the continua cannot be used for the determination of the canonical field equations of thermodynamics, i.e. the canonical field equations of thermodynamics cannot be derived in analogy to the Hamilton principle of the field theory [73].

8. Thermodynamics in Canonical Form

a) The Canonical Field Equations. The derivation of the canonical field equations of thermodynamics is due to Verhás [78] and Vojta [79] simultaneously and independently from each other. They have started by confining themselves rigorously to the mathematical analysis of the dissipation integral (6.148). They accepted that the time derivatives $\dot{\Gamma}_i$ are neutral, because in the dissipative Lagrangian density

$$\mathscr{L} = \varrho \dot{s} - \Psi = \varrho \sum_{i,k=1}^{f} s_{ik}^{-1} \dot{\Gamma}_i \dot{\Gamma}_k - \frac{1}{2} \sum_{i,k=1}^{f} L_{ik} \, \nabla \Gamma_i \cdot \nabla \Gamma_k \tag{6.149}$$

they must be considered as constant parameters in the course of variation. From this follows that the thermodynamic Lagrangian density can be given in the case of f independent scalar Γ_i parameters as

$$\mathscr{L} = \mathscr{L}(\Gamma_1, \Gamma_2, \ldots, \Gamma_f, \nabla \Gamma_1, \nabla \Gamma_2, \ldots, \nabla \Gamma_f). \tag{6.150}$$

Comparing this Lagrangian density to the Lagrange function of a conservative scleronom point system (such as represented by (6.142) if time does not occur as an independent variable), the following analogies are valid between the corresponding quantities of point mechanics and thermodynamics of continua:

Mechanics		Thermodynamics
L	$\leftrightarrow$	$\mathscr{L}$
q_i	$\leftrightarrow$	Γ_i
$\dot{q}_i$	$\leftrightarrow$	$\nabla \Gamma_i$.

It should be stressed that our refering to these analogies to mechanics serve only to further the understanding and evidently do not indicate any additional intrinsic relation between the two disciplines. Nevertheless, if we want to base a subsequent analysis upon these analogies, it must be taken into account that by the Hamilton principle of point mechanics (6.141) requires the stationarity of a time integral, whereas in the case of the variational principle of thermodynamics (6.52) the stationarity of a volume integral must be considered. Hence it is evident that the function of the operator $\frac{\mathrm{d}}{\mathrm{d}t}$ used in point mechanics is replaced in thermodynamics by that of the operator ∇ determined by the spatial derivatives $\left\{\frac{\partial}{\partial x_1}, \frac{\partial}{\partial x_2}, \frac{\partial}{\partial x_3}\right\}$. The change of rôle of the operators, i.e. the correctness of the correspondance

$$\frac{\mathrm{d}}{\mathrm{d}t} \leftrightarrow \nabla$$

can immediately be read off from the comparison of the Lagrange equations (6.143) and (6.147) belonging to the two different variational principles.

Now let us consider the independent scalar field quantities Γ_i as general coordinates and—following VOJTA and VERHÁS—let us define the general thermodynamic impulse as

$$\boldsymbol{\Pi}_i = \frac{\partial \mathscr{L}}{\partial \nabla \Gamma_i}, \quad \text{i.e.} \quad \Pi_{i\alpha} = \frac{\partial \mathscr{L}}{\partial \left(\frac{\partial \Gamma_i}{\partial x_\alpha}\right)}, \quad (\alpha = 1, 2, 3)$$
$$(i = 1, 2, \ldots, f). \tag{6.151}$$

It should be noted that in the case of vector processes (heat conduction, diffusion) the quantities $\boldsymbol{\Pi}_i$ are vectors, whereas in the case of tensor processes (viscous flow) they are second order tensors. In the case of the viscous flow, of course, according to (6.95) three scalar Γ_i parameters must be chosen as components of velocity, i.e. $\Gamma_\alpha \equiv -v_\alpha$, $(\alpha = 1, 2, 3)$. Therefore, with the above-defined generalized momenta, a so-called multi-dimensional Hamiltonian formalism is generated, since three scalar momenta $\Pi_{i\alpha}$ are conjugated to one generalized coordinate Γ_i. With the dissipative Lagrangian density (6.149) from (6.151) we have

$$\boldsymbol{\Pi}_i = -\sum_{k=1}^{f} L_{ik} \nabla \Gamma_k, \quad \text{i.e.} \quad \Pi_{i\alpha} = -\sum_{k=1}^{f} L_{ik} \frac{\partial \Gamma_k}{\partial x_\alpha}, \quad (\alpha = 1, 2, 3)$$
$$(i = 1, 2, \ldots, f). \tag{6.152}$$

Owing to the fact that the gradients $\nabla \Gamma_k$ are the dissipative forces, the general impulse vectors $\boldsymbol{\Pi}_i$ (due to the linear kinematical constitutive equations) are equal to the negatives of the current densities of the transport processes: $\boldsymbol{\Pi}_i = -\boldsymbol{J}_i$. The Euler-Lagrange field equations (6.54) [or (6.147)] of thermodynamics can again be given by making use of the definitions of general impulses (6.151) as

$$\frac{\partial \mathscr{L}}{\partial \Gamma_i} = \nabla \cdot \boldsymbol{\Pi}_i, \quad (i = 1, 2, \ldots, f). \tag{6.153}$$

Now forming the total variation of the Lagrangian density (6.150) and by using (6.151) and (6.153), the expression

$$\delta \mathscr{L} = \sum_{i=1}^{f} \left(\frac{\partial \mathscr{L}}{\partial \Gamma_i} \delta \Gamma_i + \frac{\partial \mathscr{L}}{\partial \nabla \Gamma_i} \cdot \delta \nabla \Gamma_i \right) = \sum_{i=1}^{f} (\nabla \cdot \boldsymbol{\Pi}_i \, \delta \Gamma_i + \boldsymbol{\Pi}_i \cdot \delta \, \nabla \Gamma_i) \tag{6.154}$$

is obtained, which can be given with the identity

$$\boldsymbol{\Pi}_i \cdot \delta \nabla \Gamma_i = \delta(\boldsymbol{\Pi}_i \cdot \nabla \Gamma_i) - \nabla \Gamma_i \cdot \delta \boldsymbol{\Pi}_i \tag{6.155}$$

also in the following form:

$$\delta \left(\sum_{i=1}^{f} \boldsymbol{\Pi}_i \cdot \nabla \Gamma_i - \mathscr{L} \right) = \sum_{i=1}^{f} (\nabla \Gamma_i \cdot \delta \boldsymbol{\Pi}_i - \nabla \cdot \boldsymbol{\Pi}_i \, \delta \Gamma_i) . \tag{6.156}$$

From this it can be seen that on the left-hand side the total variation of a function

$$\mathscr{H} \equiv \sum_{i=1}^{f} \boldsymbol{\Pi}_i \cdot \nabla \Gamma_i - \mathscr{L} \tag{6.157}$$

is found, from which the gradients $\nabla \Gamma_i$ can be expressed with (6.151) by the variables $\boldsymbol{\Pi}_i$. Proceeding in this manner, a new function

$$\mathscr{H} = \mathscr{H}(\Gamma_1, \Gamma_2, \ldots, \Gamma_f, \boldsymbol{\Pi}_1, \boldsymbol{\Pi}_2, \ldots, \boldsymbol{\Pi}_f) \tag{6.158}$$

can be introduced which is a function of the general coordinates Γ_i and the general impulse vectors $\boldsymbol{\Pi}_i$. This function can be called the *dissipative or thermodynamical Hamiltonian density.*

On the basis of the foregoing the canonical field equations of thermodynamics can be given immediately. Forming the total variation of the dissipative Hamiltonian density

$$\delta \mathscr{H} = \sum_{i=1}^{f} \left(\frac{\partial \mathscr{H}}{\partial \Gamma_i} \delta \Gamma_i + \frac{\partial \mathscr{H}}{\partial \boldsymbol{\Pi}_i} \cdot \delta \boldsymbol{\Pi}_i \right) \tag{6.159}$$

with (6.156), the following equations are obtained:

$$\frac{\partial \mathscr{H}}{\partial \boldsymbol{\Pi}_i} = \nabla \Gamma_i \quad (i = 1, 2, \ldots, f) \tag{6.160a}$$

and

$$\frac{\partial \mathscr{H}}{\partial \Gamma_i} = -\nabla \cdot \boldsymbol{\Pi}_i, \ (i = 1, 2, \ldots, f) . \tag{6.160b}$$

These are the canonical field equations of thermodynamics. It is evident that the first group of the equations represents f vector equations, whereas the second group represents f scalar equations as a consequence of the multi-dimensional formalism. Otherwise, the two groups of equations are equivalent to the Euler-Lagrange field equations of thermodynamics, i.e. to the differential equations governing the irreversible transport processes. In order to demonstrate this, the function

$\mathscr{H}$ must be transformed so that it corresponds to the set of functions (6.158), i.e. where $\mathscr{H}$ is expressed in terms of the Γ_i and $\boldsymbol{\Pi}_i$ independent variables.

In the linear theory the required expression of the function $\mathscr{H}$ can be given in a simple way. In such cases the dissipation potential is a homogeneous quadratic function of the gradients and since for such functions the Euler theorem is valid, further by using (6.149) and (6.151) it can be given that

$$2\Psi = \sum_{i=1}^{f} \frac{\partial \Psi}{\partial \nabla \Gamma_i} \cdot \nabla \Gamma_i = - \sum_{i=1}^{f} \frac{\partial \mathscr{L}}{\partial \nabla \Gamma_i} \cdot \nabla \Gamma_i = - \sum_{i=1}^{f} \boldsymbol{\Pi}_i \cdot \nabla \Gamma_i \quad (6.161)$$

by which from (6.157) we have

$$\mathscr{H} = - \varrho \dot{s} - \Psi. \quad (6.162)$$

Here the function $\mathscr{H}$ is a function of the general coordinates Γ_i, included in $\dot{s}$, and of the general velocities $\nabla \Gamma_i$, incorporated in Ψ, i.e. $\mathscr{H} = \mathscr{H}(\Gamma_i, \nabla \Gamma_i)$. Therefore, the $\mathscr{H}$ function in (6.162) does not yet correspond to the dissipative Hamiltonian density (6.158), which is necessary to given the actual forms of the canonical field equations (6.160). In linear irreversible thermodynamics, however, the dissipation potential Ψ may be replaced by a homogeneous quadratic function of the current densities (general impulses), i.e. by the

$$\Phi \equiv \frac{1}{2} \sum_{i,k=1}^{f} R_{ik} \boldsymbol{\Pi}_i \cdot \boldsymbol{\Pi}_k \quad (6.163)$$

dissipation potential. Therefore, instead of (6.162) the function $\mathscr{H}$ can also be given as

$$\mathscr{H} = - \varrho \dot{s} - \Phi, \quad (6.164)$$

which already satisfies the expected requirements and which can be called the *dissipative Hamiltonian density*.

With the dissipation Hamiltonian density (6.164), from the first group of the canonical equations (6.160 a) the Onsager linear constitutive equations are as follows:

$$- \sum_{k=1}^{f} R_{ik} \boldsymbol{\Pi}_k = \nabla \Gamma_i, \quad (i = 1, 2, \ldots, f). \quad (6.165)$$

These are also the alternative forms of the general impulse definitions (6.152). From the second group of the canonical equations (6.160 b) the following field equations are obtained:

$$\frac{\partial \varrho \dot{s}}{\partial \Gamma_i} = \nabla \cdot \boldsymbol{\Pi}_i, \quad (i = 1, 2, \ldots, f) \quad (6.166)$$

which are, considering the expression (6.122), clearly identical to the wellknown balance equations. Thus, for instance, in the case of heat conduction in solid bodies, when

$$\varrho\dot{s} = \frac{\varrho}{T}\frac{\partial u}{\partial t}, \quad \Gamma_q \equiv \frac{1}{T}, \quad \boldsymbol{\Pi}_q \equiv -\boldsymbol{J}_q, \quad \Phi \equiv \frac{R_{qq}}{2}\boldsymbol{J}_q^2 = \frac{L_{qq}^{-1}}{2}\boldsymbol{J}_q^2$$

the first group of the canonical equations (6.160a) yields the linear law

$$R_{qq}\boldsymbol{J}_q = \nabla\left(\frac{1}{T}\right), \text{ i.e. } \boldsymbol{J}_q = L_{qq}\,\nabla\left(\frac{1}{T}\right)$$

whereas the second group leads to the equation

$$\varrho\frac{\partial u}{\partial t} + \nabla \cdot \boldsymbol{J}_q = 0$$

which is the energy balance.[1] It is evident from the above treatment that in the theory of non-equilibrium thermodynamics we can construct a Hamiltonian formalism in multi-dimensional and local form, which is equivalent to the Lagrangian formalism represented by the expressions (6.51) to (6.54) and, as general. This fact is a plausible consequence of the existence of the integral principle.

b) Legendre Transformations. In the Gibbsian theory of the equilibrium potential functions [81], further in analytical mechanics the dual transformation recognized by Legendre plays a very important rôle [43]. Thus it as might have been expexted that with the formulation of the non-equilibrium thermodynamics in canonical forms, the Legendre transformations are of fundamental importance here as well.

Let us consider a function of f independent variables $x_1, x_2, \ldots, x_f$, i.e.

$$\Psi = \Psi(x_1, x_2, \ldots, x_f) \tag{6.167}$$

the total variation of which is

$$\delta\Psi = \sum_{i=1}^{f}\frac{\partial\Psi}{\partial x_i}\,\delta x_i = \sum_{i=1}^{f} y_i\,\delta x_i \tag{6.168}$$

where by means of the transformation

$$\frac{\partial\Psi}{\partial x_i} = y_i, \qquad (i = 1, 2, \ldots, f) \tag{6.169}$$

the new variables $y_1, y_2, \ldots, y_f$ have been introduced. In this way the number of variables has been doubled. Let us define a new function Ψ_λ, which is called the Legendre transform of Ψ, in the following manner:

$$\Psi_\lambda(y_1, y_2, \ldots, y_\lambda, x_{\lambda+1}, x_{\lambda+2}, \ldots, x_f) = \sum_{i=1}^{\lambda} y_i x_i - \Psi. \tag{6.170}$$

[1] Of course, we can construct the canonical equations in different pictures too [79, 80].

The index λ here indicates that the function Ψ_λ is expressed in terms of the old (or passive) variables which do not participate in the transformation $x_{\lambda+1}, x_{\lambda+2}, \ldots, x_\lambda$, and in terms of the new variables $y_1, y_2, \ldots, y_\lambda$. Let us consider the total variation of Ψ_λ:

$$\delta\Psi_\lambda = \sum_{i=1}^{\lambda} (y_i\,\delta x_i + x_i\,\delta y_i) - \delta\Psi = \sum_{i=1}^{\lambda} x_i\,\delta y_i - \sum_{i=\lambda+1}^{f} y_i\,\delta x_i, \qquad (6.171)$$

where (6.168) and (6.169) have been used. If $\lambda = f$, i.e. if each variables is active, we have

$$\delta\,\Psi_f = \sum_{i=1}^{f} x_i\,\delta y_i \qquad (6.172)$$

which can be considered as the total variation of such a function $\Phi \equiv \Psi_f$, i.e.

$$\Phi = \Phi(y_1, y_2, \ldots, y_f) \qquad (6.173)$$

for which

$$\delta\Phi = \sum_{i=1}^{f} \frac{\partial\Phi}{\partial y_i}\,\delta y_i \qquad (6.174)$$

and

$$\frac{\partial\Phi}{\partial y_i} = x_i, \quad (i = 1, 2, \ldots, f) \qquad (6.175)$$

are valid. Hence it can be said that Φ is the total Legendre transformed function of Ψ. This statement, however, is also valid in a reverse sense, since by the transformations (6.169) and (6.175) a total symmetry is demonstrated in the variables $x_1, x_2, \ldots, x_f$ and $y_1, y_2, \ldots, y_f$, further in the functions Ψ and Φ, respectively. Owing to this duality, the Legendre transformation is often called *dual transformation.*

Let us apply the foregoing to thermodynamics. In case of the choice $x_i \equiv \nabla\Gamma_i$ and $y_i \equiv \boldsymbol{\Pi}_i$, however, it can be seen immediately that the dissipation potential Φ is the total Legendre transform of the dissipation potential Ψ, and conversely. Hence for the dissipation potentials

$$\Psi = \Psi(\nabla\Gamma_1, \nabla\Gamma_2, \ldots, \nabla\Gamma_f) \text{ and } \Phi = \Phi(\boldsymbol{\Pi}_1, \boldsymbol{\Pi}_2, \ldots, \boldsymbol{\Pi}_f)$$

the Legendre dual transformations are valid:

$$\frac{\partial\Phi}{\partial\nabla\Gamma_i} = \boldsymbol{\Pi}_i, \qquad \frac{\partial\Phi}{\partial\boldsymbol{\Pi}_i} = \nabla\Gamma_i \qquad (i = 1, 2, \ldots, f), \qquad (6.176)$$

and

$$\Psi = \sum_{i=1}^{f} \boldsymbol{\Pi}_i \cdot \nabla\Gamma_i - \Phi, \qquad \Phi = \sum_{i=1}^{f} \boldsymbol{\Pi}_i \cdot \nabla\Gamma_i - \Psi.$$

It should be noted that in the case of the linear Onsager theory the first two sets of equations are identical to the linear kinematical equations [see (4.11) and (4.12)], because in such cases Ψ and Φ are homogeneous quadratic functions of the independent variables. Of course, in general (non-linear) cases also the Legendre transformation creates a relation between the dissipation potentials, and consequently, the given formalism can be applied in each non-linear theory, in which the dissipation potentials may be defined.

On the basis of the foregoing it is evident that when we are changing from the expression (6.162) of the dissipative Hamiltonian density to the expression (6.164), a Legendre transformation has been applied implicitly. It can easily be seen that from the Lagrangian density (6.150) the Hamiltonian density (6.158) can also be obtained by means of a total Legendre transformation of the variables $\nabla \Gamma_f$ and $\boldsymbol{\Pi}_i$ which is represented by (6.157). The variables $\Gamma_1, \Gamma_2, \ldots, \Gamma_i$ do not participate in this transformation, so that the general coordinates are passive variables. With the total variation of the function $\mathscr{H}$ from (6.157) and (6.158), we get

$$\delta \mathscr{H} = \sum_{i=1}^{f} \delta(\nabla \Gamma_i \cdot \boldsymbol{\Pi}_i) - \delta \mathscr{L} = \sum_{i=1}^{f} \nabla \Gamma_i \cdot \delta \boldsymbol{\Pi}_i - \sum_{i=1}^{f} \frac{\partial \mathscr{L}}{\partial \Gamma_i} \delta \Gamma_i$$

$$\delta \mathscr{H} = \sum_{i=1}^{f} \frac{\partial \mathscr{H}}{\partial \Gamma_i} \delta \Gamma_i + \sum_{i=1}^{f} \frac{\partial \mathscr{H}}{\partial \boldsymbol{\Pi}_i} \cdot \delta \boldsymbol{\Pi}_i = \sum_{i=1}^{f} \frac{\partial \mathscr{H}}{\partial \Gamma_i} \delta \Gamma_i + \sum_{i=1}^{f} \nabla \Gamma_i \cdot \delta \boldsymbol{\Pi}_i$$

respectively. From the comparison of which the very important relation

$$\frac{\partial \mathscr{L}}{\partial \Gamma_i} = -\frac{\partial \mathscr{H}}{\partial \Gamma_i}, \qquad (i = 1, 2, \ldots, f) \tag{6.177}$$

is obtained for the derivatives with respect to the passive variables. These results can be summarized in the following way.

Applying the Legendre transformation to the Lagrangian density $\mathscr{L}$ so that the active variables of the transformation are the general velocities $\nabla \Gamma_i$ whereas the passive variables are the general coordinates Γ_i, the general velocities are transformed into the general impulses $\boldsymbol{\Pi}_i$ and the Lagrangian density

$$\mathscr{L} = \mathscr{L}(\Gamma_1, \Gamma_2, \ldots, \Gamma_f, \nabla \Gamma_1, \nabla \Gamma_2, \ldots, \nabla \Gamma_f) = \varrho \dot{s} - \Psi \tag{6.178}$$

is transformed into the

$$\mathscr{H} = \mathscr{H}(\Gamma_1, \Gamma_2, \ldots, \Gamma_f, \boldsymbol{\Pi}_1, \boldsymbol{\Pi}_2, \ldots, \boldsymbol{\Pi}_f) = -\varrho \dot{s} - \Phi \tag{6.179}$$

the Hamilton density.

c) The Canonical Form of the Dissipative Integral. Now let us consider, because of the duality of the Legendre transformation, the following Lagrangian density

$$\mathscr{L} = \sum_{i=1}^{f} \boldsymbol{\Pi}_i \cdot \nabla \Gamma_i - \mathscr{H}(\Gamma_1, \Gamma_2, \ldots, \Gamma_f, \boldsymbol{\Pi}_1, \boldsymbol{\Pi}_2, \ldots, \boldsymbol{\Pi}_f). \tag{6.180}$$

Its variation with respect to the $\boldsymbol{\Pi}_i$ impulses vanishes, i.e.

$$\delta_{\boldsymbol{\Pi}_i}\mathscr{L} = \sum_{i=1}^{f}\left(\nabla\Gamma_i - \frac{\partial\mathscr{H}}{\partial\boldsymbol{\Pi}_i}\right)\cdot\delta\boldsymbol{\Pi}_i = 0 \tag{6.181}$$

since the coefficient of $\delta\boldsymbol{\Pi}_i$ is zero according to (6.160a). This means that the variation of $\boldsymbol{\Pi}_i$ does not influence the variation of $\mathscr{L}$, and therefore the general impulses in the integrand of the dissipative integral can be considered as independent variables. In other words, the dissipation integral can be given without any change in the Lagrangian density (6.180) as

$$L = \int_V\left(\sum_{i=1}^{f}\boldsymbol{\Pi}_i\cdot\nabla\Gamma_i - \mathscr{H}(\Gamma_1, \Gamma_2, \ldots, \Gamma_f, \boldsymbol{\Pi}_1, \boldsymbol{\Pi}_2, \ldots, \boldsymbol{\Pi}_f)\right)\mathrm{d}V = \max \tag{6.182}$$

which is valid in the case of the arbitrary but independent variation of the canonical conjugated Γ_i and $\boldsymbol{\Pi}_i$ variables. This integral can be called *the canonical form of the dissipative integral* corresponding to its wellknown mechanical equivalent [44].

The canonical form of the dissipative integral means, of course, a new variational task, to which the following Euler-Lagrange equations belong:

$$\frac{\partial\mathscr{L}}{\partial\boldsymbol{\Pi}_i} - \sum_{\alpha=1}^{3}\frac{\partial}{\partial x_\alpha}\frac{\partial\mathscr{L}}{\partial\left(\dfrac{\partial\boldsymbol{\Pi}_i}{\partial x_\alpha}\right)} = \left(\nabla\Gamma_i - \frac{\partial\mathscr{H}}{\partial\boldsymbol{\Pi}_i}\right) - \sum_{\alpha=1}^{3}\frac{\partial}{\partial x_\alpha}\cdot 0 = 0 \tag{6.183a}$$

and

$$\frac{\partial\mathscr{L}}{\partial\Gamma_i} - \sum_{\alpha=1}^{3}\frac{\partial}{\partial x_\alpha}\frac{\partial\mathscr{L}}{\partial\left(\dfrac{\partial\Gamma_i}{\partial x_\alpha}\right)} = -\frac{\partial\mathscr{H}}{\partial\Gamma_i} - \sum_{\alpha=1}^{3}\frac{\partial\Pi_{i\alpha}}{\partial x_\alpha} = 0. \tag{6.183b}$$

which are exactly identical to the canonical field equations (6.160a) and (6.160b).

9. Conclusions

In the possession of the developed theory we can easily be convinced that the canonical transformations, the Poisson brackets etc. are also valid and applicable in the theory of dissipative processes. Therefore only the problem referring to the future development of thermodynamics will be briefly mentioned, which is related to the development of a non-linear theory.

The "kinetic" term $\varrho\dot{s}$ of the dissipative Lagrangian $\mathscr{L}$ and Hamiltonian $\mathscr{H}$ densities has a general validity, and consequently it also remains valid for the case of non-linear problems. This is absolutly consistent as far as entropy can be regarded an unique macroparameter in the sense of the local (cellular) equilibrium. Therefore, the development of a non-linear thermodynamic theory depends primarily on whether the potential terms of the functions $\mathscr{L}$ and $\mathscr{H}$, i.e. the potentials Ψ and Φ could be given more generally as the homogeneous quadratic expression used in linear theory. In this respect, the dissipation potentials given in Chapter V. sect. 5. are relevant, even if the confirmation of the second order reciprocal relations ensuring their potential character is heretofore not yet satisfactory in every respect. It must be acknowledged that in the absence of dissipative potentials neither linear nor non-linear theories do exist. Completing this theoretical requirement by the practical one viz. that our non-linear theory should not be too complicated, not much choice remains between the dissipative potentials. It is very unfortunate that sometimes non-linear interactions are preferred by nature, whereas physicists do not like non-linear equations. Therefore, the development of non-linear thermodynamics will be doubtlessly the result of a compromise.

Of course, the validity of the canonical formalism of thermodynamics will not be influenced by the non-linear theory to be developed. Only the actual first group (6.165) of the canonical field equations will be modified according to the dissipative potential chosen as fundamental. The integral principle and the second group of the canonical field equations (balance equations) remain valid in every case. This must be so, since the embracing capacity and accuracy of the canonical formalism worked out by Euler, Lagrange and Hamilton is based upon the mathematical methods of the variational calculus and is independent of the physical discipline chosen as subject of application. It is another question, that this immensely powerful weapon of mathematical physics has only been applied in thermodynamics these days, though the famous works of LAGRANGE (Mécanique Analytique 1788), FOURIER (Theorie Analytique de la Châleur 1822), NAVIER (1822), STOKES (1845) and FICK (Über Diffusion 1855) furnished sufficient basis long ago already. This delay of as much as a century was the reason for preparing this book on the principle "bis dat qui cito dat". So the author asks the reader to judge the imperfectness of the book—particularly Chapter VI—taking the unsuccessful research of a century into consideration.

Appendix

On the Elements of Vector and Tensor Calculus

For convenience of the reader let us briefly summarize the fundamental elements of the vector and tensor calculus. The exclusive aim, however, of this summary is to give the used notations and denominations.

In the three-dimensional Euclidean space a physical quantity which can be represented by a tensor of ν-th order has 3^ν components, a scalar quantity can be considered as a tensor of zero order, whereas a vector as a tensor of first order. The letter used to denote a scalar quantity was always printed in *italics*: a scalar, (tensor of zero order), that used to denote an arbitrary vector by ***italic bold-face*** letter-type: $\boldsymbol{v}$ vector, its components v_α, $(\alpha = 1, 2, 3)$ (tensor of the first order), whereas the tensor of the second or higher order is denoted by sans-serif-type. For example: T second order tensor, its components $T_{\alpha\beta}$, $(\alpha, \beta = 1, 2, 3)$.

In the following, the operations between the quantities of different tensorial order are given in a symbolic form, then after the sign $\leftrightarrow$ (referring to the equivalency) we give them with components as well. We always use a right-handed Cartesian coordinate system. On the other hand, the convention used in tensor calculus is accepted, according to which we are automatically summing with respect to the indices occurring twice (or many times) in an expression. Hence, for instance we write

$$v_\alpha = T_{\alpha\beta} w_\beta \qquad (\alpha, \beta = 1, 2, 3)$$

instead of

$$v_\alpha = \sum_{\beta=1}^{3} T_{\alpha\beta} w_\beta \qquad (\alpha = 1, 2, 3)$$

with the usual convention that a repeated index is summed from 1 to 3. Similarly, we do not write the running indices, since these always take the values 1, 2, 3, and denote the Cartesian coordinates of the tensor quantities in question.

1. Fundamental Concepts and Simple Operations

Owing to the fact that the scalars and vectors are particular tensor quantities, it is sufficient to summarize only the algebra of the latter ones.

The unit tensor $\boldsymbol{\delta}$ is interpreted by the matrix of $[\delta_{\alpha\beta}]$, i.e.:

$$\boldsymbol{\delta} \leftrightarrow [\delta_{\alpha\beta}] \tag{1}$$

where $\delta_{\alpha\beta}$ is the Kronecker delta function, the value of which is

$$\delta_{\alpha\beta} = \begin{cases} 0 & \text{if } \alpha \neq \beta, \\ 1 & \text{if } \alpha = \beta. \end{cases} \tag{2}$$

Two R and S tensors are equal

$$\mathsf{R} = \mathsf{S} \quad \text{if} \quad R_{\alpha\beta} = S_{\alpha\beta} \tag{3}$$

all their elements are equal, i.e. if their matrix is equal in the sense

$$[R_{\alpha\beta}] = [S_{\alpha\beta}]. \tag{4}$$

Under the sum (difference) of two tensors **R** and **S** the tensor **T** is to be understood, whose components are determined by the sum (difference) of the components of the tensors **R** and **S**. Hence,

$$\mathbf{T} = \mathbf{R} \pm \mathbf{S} \leftrightarrow T_{\alpha\beta} = R_{\alpha\beta} \pm S_{\alpha\beta} \tag{5}$$

or in matrix form

$$[T_{\alpha\beta}] = [R_{\alpha\beta}] \pm [T_{\alpha\beta}]. \tag{6}$$

The product of a tensor **T** and a scalar a determines a tensor **S**, whose components are a times the corresponding components of **T**, i.e.

$$\mathbf{S} = a\mathbf{T} \leftrightarrow S_{\alpha\beta} = aT_{\alpha\beta}. \tag{7}$$

In the following, the matrix formalism will not be given, but it should be noted that the operations between second order tensors can always be carried out with the aid of the corresponding matrix operations.

2. Symmetric and Antisymmetric Tensors

If the components of a tensor **T** determined by the indices are interchanged, the transposed tensor is obtained:

$$\widetilde{T}_{\alpha\beta} = T_{\beta\alpha}. \tag{8}$$

A tensor **T** is symmetric if

$$\mathbf{T} = \widetilde{\mathbf{T}} \leftrightarrow T_{\alpha\beta} = T_{\beta\alpha} \tag{9}$$

and antisymmetric if

$$\mathbf{T} = -\widetilde{\mathbf{T}} \leftrightarrow T_{\alpha\beta} = -T_{\beta\alpha}. \tag{10}$$

It follows from definitions (9) and (10) that a second order symmetric tensor has six, whereas an antisymmetric tensor has three independent components. Since the components of any tensor can be given in the form

$$T_{\alpha\beta} = \frac{1}{2}(T_{\alpha\beta} + T_{\beta\alpha}) + \frac{1}{2}(T_{\alpha\beta} - T_{\beta\alpha}) \tag{11}$$

it is evident that every arbitrary asymmetric tensor may be split up into a symmetric

$$\mathbf{T}^{s} = \frac{1}{2}(\mathbf{T} + \widetilde{\mathbf{T}}) \leftrightarrow T^{s}_{\alpha\beta} = \frac{1}{2}(T_{\alpha\beta} + T_{\beta\alpha}) \tag{12}$$

and an antisymmetric part

$$\mathbf{T}^{a} = \frac{1}{2}(\mathbf{T} - \widetilde{\mathbf{T}}) \leftrightarrow T^{a}_{\alpha\beta} = \frac{1}{2}(T_{\alpha\beta} - T_{\beta\alpha}), \tag{13}$$

i.e.

$$\mathbf{T} = \mathbf{T}^s + \mathbf{T}^a \tag{14}$$

which is expressed in components by (11).

3. Tensor Products

The exterior product of two tensors of arbitrary order ν and μ gives a tensor of order $\mu + \nu$. Thus, if $\boldsymbol{v}$ and $\boldsymbol{w}$ are two vectors, the particular second order tensor determined by the exterior product

$$\boldsymbol{vw} \leftrightarrow (\boldsymbol{vw})_{\alpha\beta} = v_\alpha w_\beta \tag{15}$$

is called a dyad

$$\mathbf{D} \equiv \boldsymbol{vw} \leftrightarrow \mathrm{D}_{\alpha\beta} = v_\alpha w_\beta, \tag{16}$$

whereas the exterior product in question is called the dyadic product of two vectors. The dyadic product of two vectors, a dyad, is a second order tensor, but in general, a second order tensor cannot be represented by the dyadic product of two vectors. In the three-dimensional Euclidean space, however, the general tensor formalism and the dyad formalism are equivalent to each other. Therefore, what has been said in the preceding part concerning the symmetry properties of the tensors is valid also for dyads. In analogy to (8) the transposed dyad $\widetilde{\mathbf{D}}$ is, by definition

$$\widetilde{\boldsymbol{vw}} = \boldsymbol{wv} \leftrightarrow \widetilde{(vw)}_{\alpha\beta} = v_\beta w_\alpha \tag{17}$$

with the use of which it can be seen that, every dyad $\mathbf{D} \equiv \boldsymbol{vw}$ may always be split up into a symmetric

$$\mathbf{D}^s = \frac{1}{2}(\boldsymbol{vw} + \boldsymbol{wv}) \leftrightarrow \mathrm{D}^s_{\alpha\beta} = \frac{1}{2}(v_\alpha w_\beta + w_\alpha v_\beta) \tag{18}$$

and an antisymmetric part

$$\mathbf{D}^a = \frac{1}{2}(\boldsymbol{vw} - \boldsymbol{wv}) \leftrightarrow \mathrm{D}^a_{\alpha\beta} = \frac{1}{2}(v_\alpha w_\beta - w_\alpha v_\beta). \tag{19}$$

In the three-dimensional Euclidean space $\mathbf{D}^a$ has an intrinsic relation with the vector product of two ordinary (polar) vectors. If $\boldsymbol{v}$ and $\boldsymbol{w}$ are two polar vectors, their vector product, defined in the usual sense

$$\boldsymbol{C}^a = \boldsymbol{v} \times \boldsymbol{w} \leftrightarrow C^a_\gamma = v_\alpha w_\beta - w_\alpha v_\beta, \quad (\alpha, \beta, \gamma \text{ cycl.}) \tag{20}$$

represents an axial vector $\boldsymbol{C}^a$. As can be seen by comparing (19) and (20) $\boldsymbol{C}^a$ is related to the antisymmetric dyad $\mathbf{D}^a$ as follows

$$\boldsymbol{C}^a = 2\mathbf{D}^a \leftrightarrow C^a_\gamma = 2\mathrm{D}^a_{\alpha\beta}, \quad (\text{cycl.}). \tag{21}$$

Since an axial vector can be formed not only from an antisymmetric dyad but also from any second order antisymmetric tensor similar to (19):

$$T^{a}_{\gamma} = \frac{1}{2}(T_{\alpha\beta} - T_{\beta\gamma}), \qquad \text{(cycl.)} \tag{22}$$

all the quantities which can be represented by antisymmetric tensors (particularly dyads) can be expressed by axial vectors too.

From among the exterior products the exterior products of a vector with a second order tensor is used often. These are in different order of succession

$$\boldsymbol{v}\mathbf{T} \leftrightarrow (\boldsymbol{v}\mathbf{T})_{\alpha\beta\gamma} = v_{\alpha}T_{\beta\gamma} \tag{23}$$

and

$$\mathbf{T}\boldsymbol{v} \leftrightarrow (\mathbf{T}\boldsymbol{v})_{\alpha\beta\gamma} = T_{\alpha\beta}v_{\gamma} \tag{24}$$

producing a third order tensor, according to the general definition of exterior tensor products.

Now, let us define the interior (or contracted) product of two tensors, which are often called scalar products. The interior product of tensors of arbitrary order can be obtained from the exterior product by putting two neighbouring indices equal and automatically summing for these identical indices. Each interior product is denoted by inserting a dot (·) between the tensor symbols. (Therefore, particularly in English literature, the interior product is sometimes called "dot product".) Proceeding in this way, the following interior products are obtained from the exterior products (16), (23) and (24]):

$$\boldsymbol{v} \cdot \boldsymbol{w} \leftrightarrow v_{\alpha}w_{\alpha}, \tag{25}$$

$$\boldsymbol{v} \cdot \mathbf{T} \leftrightarrow (\boldsymbol{v} \cdot \mathbf{T})_{\gamma} = v_{\alpha}T_{\alpha\gamma}, \tag{26}$$

$$\mathbf{T} \cdot \boldsymbol{v} \leftrightarrow (\mathbf{T} \cdot \boldsymbol{v})_{\gamma} = T_{\gamma\alpha}v_{\alpha}. \tag{27}$$

In (25) the interior product of the two vector quantities has been determined, which was called the scalar product of two vectors. This denomination is appropriate, since the result of the product is a scalar determined by the product sums of the corresponding vector components. From the definition of the given interior products, the general rule can be read off: the interior (or dot) product of two arbitrary tensors of order μ and ν always gives a tensor of order $(\mu + \nu - 2)$. It should be noted that in general

$$\boldsymbol{v} \cdot \mathbf{T} \neq \mathbf{T} \cdot \boldsymbol{v}, \tag{28}$$

i.e. the commutative law of the multiplication does not hold. If the foregoings are extended to the interior product of two second order tensors, the definition

$$\mathbf{S}\cdot\mathbf{T} \leftrightarrow (\mathbf{S}\cdot\mathbf{T})_{\alpha\gamma} = S_{\alpha\beta}T_{\beta\gamma} \tag{29}$$

is obtained. However, in the case of two second order tensors the exterior product of two tensors may be contracted twice which is indicated by inserting a double dot (:) between the tensor symbols. By this interior product

$$\mathbf{S}:\mathbf{T} = \mathbf{T}:\mathbf{S} \leftrightarrow (\mathbf{S}:\mathbf{T}) = S_{\alpha\beta}T_{\beta\alpha} \tag{30}$$

a scalar is determined, if $\mathbf{S}$ and $\mathbf{T}$ are second order tensors. Therefore (30) is also called the scalar product of two second order tensors. In general, however, the tensorial order of a quantity determined by the "double dot product" is $(\mu + \nu - 4)$ as a result of double contraction. It should be noted that instead of (30) particularly in German literatur the definition

$$\mathbf{S}:\mathbf{T} = S_{\alpha\beta}T_{\alpha\beta} \tag{31}$$

is used often, because the choice is merely one of convention. We shall select the definition (30), which is preferred in French literature and which is also widespread in the scools of the Benelux states playing a leading rôle in thermodynamics.

The trace (spur) of a tensor $\mathbf{T}$, i.e. the invariant scalar determined as the sum of the diagonal elements of the tensor $\mathbf{T}$ can be defined as

$$\text{spur}\,\mathbf{T} = \mathbf{T}:\boldsymbol{\delta} = T_{\alpha\alpha}. \tag{32}$$

4. Tensor Derivatives

Now, let us give the most important spatial derivatives with the aid of the nabla operator:

$$\nabla \equiv \frac{\partial}{\partial \boldsymbol{r}} \leftrightarrow \frac{\partial}{\partial x_\alpha}. \tag{33}$$

The following expressions define the fundamental spatial derivatives of tensorial quantities of various order.

Scalar gradient:

$$\nabla a = \text{grad}\, a \leftrightarrow (\nabla a)_\alpha = \frac{\partial a}{\partial x_\alpha}. \tag{34}$$

Vector gradient:

$$\nabla \boldsymbol{v} = \text{Grad}\, \boldsymbol{v} \leftrightarrow (\nabla \boldsymbol{v})_{\alpha\beta} = \frac{\partial v_\beta}{\partial x_\alpha}. \tag{35}$$

Vector rotation:

$$\nabla \times \boldsymbol{v} = \text{rot}\, \boldsymbol{v} \leftrightarrow (\nabla \times \boldsymbol{v})_\gamma = \frac{\partial v_\beta}{\partial x_\alpha} - \frac{\partial v_\alpha}{\partial x_\beta}, \quad (\alpha, \beta, \gamma \text{ cycl.}) \tag{36}$$

Vector divergency:

$$\nabla \cdot \boldsymbol{v} = \operatorname{div} \boldsymbol{v} \leftrightarrow (\nabla \cdot \boldsymbol{v}) = \frac{\partial v_\gamma}{\partial x_\gamma}. \tag{37}$$

Tensor divergency:

$$\nabla \cdot \mathbf{T} = \operatorname{Div} \mathbf{T} \leftrightarrow (\nabla \cdot \mathbf{T})_\alpha = \frac{\partial T_{\beta\alpha}}{\partial x_\beta}. \tag{38}$$

Finally the definition of the Laplace operator is:

$$\nabla \cdot \nabla \equiv \operatorname{div} \operatorname{grad} \equiv \Delta \leftrightarrow \sum_{\alpha=1}^{3} \frac{\partial^2}{\partial x_\alpha^2}. \tag{39}$$

It should be emphasized again that the aim of this Appendix is not to teach the fundamentals of tensor algebra and analysis to those who are interested in it—the exposition is evidently not suitable for this purpose—but to give a guide of the denotations and denominations used.

References

1. Truesdell, C., and R. A. Toupin: The Classical Field Theories in: Handbuch der Physik, 3. Bd., 1. Tl., Berlin/Göttingen/Heidelberg: Springer 1960, p. 226.
2. Sedov, L. J.: Introduction to the Mechanics of a continuous medium, Reading/London: Addison-Wesley 1965 (Translation from Russian).
3. De Groot, S. R., and P. Mazur: Non-equilibrium Thermodynamics, Amsterdam: North-Holland Publ. Co. 1962.
4. Meixner, J., and H. G. Reik: Thermodynamik der irreversiblen Prozesse in: Handbuch der Physik, 3. Bd., 2. Tl. Berlin/Göttingen/Heidelberg: Springer 1959, p. 413.
5. Prigogine, I., and R. Defay: Thermodynamique Chimique, Liége: Maison Desoer Editions 1950.
6. Chapman, S., and T. G. Cowling: The Mathematical Theory of Non-Uniform Gases, Cambridge: University Press 1939.
7. Prigogine, I., and P. Mazur: Physica 19 (1953) 241.
8. De Groot, S. R.: Thermodynamics of irreversible processes, Amsterdam: North-Holland Publ. Co. 1951.
9. De Donder, Th., and P. van Rysselberghe: The Thermodynamic Theory of Affinity, Stanford: 1963.
10. Gyarmati, I., and G. Schay: Magyar Tud. Akad. Kém. Oszt. Közl. Budapest 19 (1963) 459.
Gyarmati, I., and J. Sándor: Magyar Tud. Akad. Kém. Oszt. Közl. Budapest 20 (1963) 375.
Sándor, J., and G. Schay: Magyar Tud. Akad. Kém. Oszt. Közl. Budapest 22 (1964) 347.
Sándor, J., and J. Verhás: Magyar Tud. Akad. Kém. Oszt. Közl. Budapest 23 (1965) 403.
Sándor, J., and J. Verhás: Magyar Tud. Akad. Kém. Oszt. Közl. Budapest 27 (1967) 105.
Gyarmati, I., and J. Sándor: Kolloid. Zsurn. Moscow T. 28 (1966) N. 3, 366.
Gyarmati, I., and J. Sándor: Kolloid. Zsurn. Moscow T. 28 (1966) N. 3, 373.
Gyarmati, I., and J. Sándor: Kolloid. Zsurn. Moscow T. 28 (1966) N. 6, 829.
11. Cauchy, A. L.: Ex. de Math. 2, Oeuvres (2) 7 (1928) 79.
12. Bearman, R. J., and J. G. Kirkwood: J. Chem. Phys. 28 (1958) 136.
13. Fitts, D. D.: Nonequilibrium Thermodynamics, New York/Toronto/London: McGraw-Hill Book Co. 1962.
14. Truesdell, C.: Rend. Lincei (8) 22, (1957) 33 158.
15. Lin, C. C., and W. H. Reid: Turbulent Flow, Theoretical Aspects in: Handbuch der Physik. 8. Bd., 2. Tl. Berlin/Göttingen/Heidelberg: Springer 1963, p. 438.
16. Kluitenberg, G. A.: Physica 28 (1962) 217, 561.
17. Prigogine, I.: Ĕtude thermodynamique des phénomènes irréversibles (Thésis), Paris: Dunod and Liége: Desoer 1947.

18. GRAD, H.: Comm. Pure and Applied Math. 5 (1952) 455.
19. MEIXNER, J.: Z. Physik 146 (1961) 145.
20. BARANOWSKI, B., and T. ROMOTOWSKI: Bull. L'Acad. Pol. Sc. série des sciences chimiques XII, 1 (1964) 71; XII, 2 (1964) 127.
21. MAZUR, P., and I. PRIGOGINE: Mem. Acad. Roy. Belg. Cl. Sci. 23 (1952).
22. ONSAGER, L.: Phys. Rev. 37 (1931) 405; 38 (1931) 2265.
23. PRIGOGINE, I.: Introduction to Thermodynamics of Irreversible Processes, Springfield: Thomaes 1955.
24. MEIXNER, J.: Thermodynamik der irreversiblen Prozesse. Aachen: (als Sonderdruck vervielfältigt) 1954.
25. DENBIGH, K. G.: Thermodynamics of the Steady State. London: Methuen 1951.
26. HAASE, R.: Ergebn. exakt. Naturw. 22 (1952) 56.
27. HAASE, R.: Thermodynamik der irreversiblen Prozesse, Darmstadt: Dr. Dietrich Steinkopff 1963.
28. GYARMATI, I.: Introduction to Irreversible Thermodynamics (in Hungarian), Budapest: MTI. 1960.
29. GUMINSKI, K.: Thermodynamicka procesŏv nieodwracalnyck, Warszawa: Paustwowe Wydawnictwo Naukove 1962.
30. RYSSELBERGHE, P.: Thermodynamics of Irreversible Processes, Paris: Hermann, and New York/Toronto/London: Blaisdell Publ. Co. 1963.
31. GYARMATI, I.: On the Principles of Thermodynamics, (Manuscript (1957)) and dissertation, Budapest 1958.
32. GYARMATI, I.: Acta. Chim. Hung. 30 (1962) 2, 147.
33. MEIXNER, J.: Ann. Physik (5) 39 (1941) 333; Z. Phys. Chem. Abt. B 53 (1943) 235.
34. PRIGOGINE, I.: Physica 15 (1949) 272.
35. REIK, H. G.: Z. Physik 148 (1957) 156, 333.
36. CURIE, P.: Oeuvres p.127, Gauthier-Villars, Paris 1908.
37. CASIMIR, H. B. G.: Rev. Mod. Phys. 17 (1945) 343.
38. GYARMATI, I.: Period. Polytechn. 5 (1961) 3, 219; 5 (1961) 4, 321.
39. RYSSELBERGHE, P.: Journ. Chem. Phys. 36 (1962) 1329.
40. LI, J. C. M.: Journ. Chem. Phys. 28 (1958) 747.
41. PITZER, K.: Pure Appl. Chem. V. 2 (1961) 207.
42. TRUESDELL, C.: Phys. Bl. 16 (1960) 512.
43. LÁNCZOS, C.: The Variational Principles of Mechanics, Univ. of Toronto Press, Toronto 1949.
44. YOURGRAU, W., and S. MANDELSTAM: Variational Principles in Dynamics and Quantum Theory. Second Edition, Pitman and Sons, Ltd. London 1960.
45. GYARMATI, I., and J. SÁNDOR: Period. Polytechn. 6 (1962) 244; 7 (1963) 35.
46. ONSAGER, L., and S. MACHLUP: Phys. Rev. 91 (1953) 1505, 1512.
47. TISZA, L., and J. MANNING: Phys. Rev. 105 (1957) 1695.
48. ONO, S.: Adv. Chem. Phys. Vol. III (1961) 267.
49. PRIGOGINE, I.: Bull. Acad. Roy. Belg. Cl. Sci. 31 (1945) 600.
50. GYARMATI, I.: Zsurn. Fiz. Himii (Moscow) T. 39 (1965) N. 6, 1489.
51. GYARMATI, I.: Acta Chim. Hung. 43 (1965) 353.
52. GYARMATI, I.: Period. Polytechn. 9 (1965) 2, 205.
53. GYARMATI, I.: Acta Chim. Hung. 47 (1966) 63.
54. KIRKALDY, J. S.: Canad. Journ. Phys. 42 (1964) 1447.
55. GYARMATI, I.: Zeitschr. Phys. Chem. 234 (1967) 371.
56. Lord RAYLEIGH, (J. W. STRUTT): Proc. math. Soc. London 4 (1873) 357.
57. PRIGOGINE, I.: Bull. Acad. Roy. Belg. Cl. Sci. 40 (1954) 471.
58. BUDÓ, Á.: Mechanics (in Hungarian). Third Edition, Budapest 1964.

59. VERHÁS, J.: Period. Polytechn. 9 (1965) 2, 209; International Commitee of Electrochemical Thermodynamics and Kinetics (CITCE) 16-th Meeting, Budapest 1965.
60. VERHÁS, J.: The treatment of Transport Processes with Variational Principles (in Hungarian). Dissertation, Budapest 1965.
61. VERHÁS, J.: Zsurn. Fiz. Himii. (Moscow) T. 40 (1966) N. 6, 1213; Conf. on Some Aspects of Phys. Chem. V. 763, Budapest 1966.
62. LAX, M.: Rev. Mod. Phys. 32 (1960) 25.
63. GYARMATI, I., and K. OLÁH: Acta Chim. Hung. 35 (1963) 95.
64. GLANSDORFF, P., and I. PRIGOGINE: Physica 20 (1954) 773.
65. GLANSDORFF, P.: Bull. Acad. Roy. Belg. Cl. Sci. 42 (1956) 628.
66. GLANSDORFF, P.: Molecular Physics 3 (1960) 277.
67. GLANSDORFF, P., I. PRIGOGINE and D. HAYS: The Physics of Fluids 5 (1962) 144.
68. SCHECHTER, R. S.: Chemical Eng. Sci. 17 (1962) 803.
69. HAYS, D.: Bull. Acad. Roy. Belg. Cl. Sci. 49 (1963) 576.
70. GLANSDORFF, P., and I. PRIGOGINE: Physica 30 (1964) 351.
71. PRIGOGINE, I., and P. GLANSDORFF: Physica 31 (1965) 1242.
72. LI, J. C. M.: J. Chem. Phys. 33 (1962) 616; 37 (1962) 1592.
73. GYARMATI, I.: Acta Chim. Hung. 47 (1966) 367.
74. VERHÁS, J.: Zeitschr. Phys. Chem. 234 (1967) 226.
75. BÖRÖCZ, Sz.: Zeitschr. Phys. Chem. 234 (1967) 26; Conf. on Some Aspects of Phys. Chem. V. 769. Budapest 1966. The treatment of viscous processes with the Variational Principles of Thermodynamics (in Hungarian). Dissertation, Budapest 1967.
76. VERHÁS, J.: Zsurn. Fiz. Himii (Moscow) T. 40 (1966) N. 10, 2482.
77. VOJTA, G.: Abhandl. der DAW zu Berlin, Klass. Math. Phys. und Techn. Nr. 1 (1967) 183.
78. VERHÁS, J.: Ann. Phys. 7 (1967) 20, 90.
79. VOJTA, G.: Acta Chim. Hung. 54 (1967) 55.
80. FARKAS, H.: The treatment of heat conduction with Variational Principles (in Hungarian). Dissertation, Budapest 1967; Zeitsch. Phys. Chem. 239 (1968) 124.
81. SEMENCHENKO, V. K.: Selected Topics from Theoretical Physics (in Russian), Moscow 1966.
82. TERLETSKY, J. P.: Nouvo Cimento 7 (1958) 308.
83. MAGALINSKY, V. B., and J. P. TERLETSKY: Ann. Phys. 5 (1960) 296.
84. TERLETSKY, J. P.: Statistical Physics (in Russian). Moscow 1966.
85. TERLETSKY, J. P., and TANG NGUYEN: Ann. Phys. 7 (1967) 19, 5–6, 299.